Science: a second level course

S269 Earth and Life

Origins of Earth and Life

Prepared for the Course Team by Iain Gilmour, Ian Wright
and John Wright (academic editor)
Some sections partly based on material contributed by Steve Drury

The Open University

S269 Course Team

Course Team Chair Peter Francis

Book Editor John Wright

Course Team Angela Colling
Nancy Dise
Steve Drury
Peter Francis
Iain Gilmour
Nigel Harris
Allister Rees
Peter Skelton
Bob Spicer
Charles Turner
Chris Wilson
Ian Wright
John Wright

Course Managers Kevin Church
Annemarie Hedges

Secretaries Anita Chhabra
Janet Dryden
Marilyn Leggett
Jo Morris
Rita Quill
Denise Swann

Series Producer David Jackson

Editors Gerry Bearman
Rebecca Graham
Kate Richenburg

Graphic Design Sue Dobson
Ray Munns
Pam Owen
Rob Williams

Liaison Librarian John Greenwood

Course Assessor Professor W. G. Chaloner (FRS)

The Open University, Walton Hall, Milton Keynes MK7 6AA

First published 1997

Edited, designed and typeset by The Open University.

Printed in the United Kingdom by Jarrold Book Printing,
3 Fison Way, Thetford, Norfolk IP24 1HT

ISBN 0 7492 8182 0

This text forms part of an Open University Second Level Course.
If you would like a copy of *Studying with The Open University*, please write to the Course Reservations and Sales Centre, PO Box 724, The Open University, Walton Hall, Milton Keynes MK7 6ZS.
If you have not enrolled on the Course and would like to buy this or other Open University material, please write to Open University Educational Enterprises Ltd, 12 Cofferidge Close,
Stony Stratford, Milton Keynes MK11 1BY, United Kingdom.

1.1

S269b1oei1.1

Contents

Preface

This book is about origins, about the origin of the chemical elements and the Solar System, and about the possible ways in which life began on Earth, perhaps more than 4 billion years ago.

We start with an attempt to define life and the question of why Planet Earth seems to be the sole repository of life within the Solar System. We then go back to the beginning. First, we explore the way in which the Big Bang led to formation of the chemical elements by fusion from hydrogen, the original cosmic 'fuel', and eventually to the formation of our own Sun and Solar System, including the Earth and its moon, some 4.6 billion years ago. Conditions on the early Earth, soon after its formation, are difficult to determine, but the geological record and observations of other planetary bodies enable us to infer that conditions at the Earth's surface were very different from what they are today. This helps to explain why there was an hiatus of many hundred million years before the earliest life forms could develop and so begin the long process of evolution to complex multicellular organisms. Our story in this book, however, ends with the various hypotheses propounded to explain how life first came to be established on Earth, and the experiments that have been performed to test those hypotheses.

You will find questions designed to help you to develop arguments and/or test your own understanding as you read. Answers to unnumbered questions follow them within the text, while answers to numbered questions are at the back of the book. Questions following end-of-chapter summaries are intended to help you consolidate your understanding and to help with revision. Important terms and concepts are printed in **bold type** where they are introduced or defined.

This is the first of four books that, together with a Set Book, make up the Open University Course S269 *Earth and Life*. As you have read in the preceding paragraphs, this book is about origins.

The Dynamic Earth is about the Earth today, about the surface conditions that make it habitable for life, and how our climate is regulated. Life on Earth interacts with non-biological (abiotic) processes to the extent that the biosphere is not a separate entity but actually forms part of the lithosphere, hydrosphere and atmosphere systems. Physical and biogeochemical cycles redistribute materials (including nutrients and other elements) within and between these systems, and continually modify the shape and appearance of the Earth's surface.

As its title suggests, *Atmosphere, Earth and Life*, is about the evolution of the Earth's atmosphere from its initial condition, devoid of oxygen and rich in carbon dioxide, to its present breathable state, with plenty of oxygen. We attempt to establish whether the change was continuous and regular, or intermittent and variable, and also to discover the extent to which atmospheric and biological evolution are linked.

Evolving Life and the Earth continues the exploration of the way Earth and life have interacted to make the world we live in a very different one from that which would have resulted had the planet remained lifeless. This is *not* the same as saying that Earth and life have together provided a consistently self-regulating planet in the sense of a balanced and constant environment (i.e. biological homeostasis). The thesis of the book (indeed of the Course as a whole) is that the Earth *changes* and life *evolves,* though it is fair to say that life has itself brought about significant changes to the Earth's physical environment.

To conclude the Course, we examine the rapid interactions between Earth and Life taking place over the past 10 000 years, using a Set Book, *The Holocene: An Environmental History* by Neil Roberts. During this period, the Earth was emerging from the last great glaciation; climate was changing rapidly, and geography was settling down to resemble what we know today. Both plant and animal species, including early humans, were forced to migrate to keep pace with shifts in climatic belts. Such changes had taken place repeatedly over the previous million or so years of the Ice Age, driven by cyclical changes in the Earth's orbit and the orientation of its rotation axis. It is against the rapidly changing conditions of the Holocene, however, that civilization began to take shape. During this period, instead of merely sharing the externally driven environmental changes along with plants and other animals, humans also began to radically change their environment, fundamentally altering the relationship between Earth and Life. An examination of the Holocene record of change thus provides a unique perspective on the environmental changes that now concern us all.

Chapter 1
What is life?

***life**, n. ... the condition which distinguishes active animals and plants from inorganic matter, including the capacity for growth, functional activity and continual change preceding death...*

The Concise Oxford Dictionary, *9th edition (Oxford University Press, 1995)*

... all life shares the property of increasing the local order by making the environment around it more chaotic ... Life is sustained, growing imbalance, or 'disequilibrium' ... The degree of sophistication of a living organism can be measured by its degree of disequilibrium from a natural environment. Life ... cannot stay still. It must grow. If it stops, it dies and attains equilibrium as a corpse.

Euan B. Nisbet, Living Earth *(HarperCollins, 1991)*

At even the most primordial level, living seems to entail sensation, choosing, mind ... All [organic beings] are sentient ... Life is matter that chooses.

Lynn Margulis and Dorion Sagan, What is Life? *(Simon and Schuster, 1995)*

Imagine you are a visitor from some distant planet approaching the Earth. As you move closer, your view of the planet improves and you are able to see more and more detail (Figure 1.1). Does this planet support life? How can you decide?

Figure 1.1 A view of Earth from space showing North and South America.

Assume your spacecraft is equipped with sophisticated remote sensing and analytical equipment, and that your optical range is in the visible part of the spectrum (i.e. that you see colours as we ordinary mortals do). The occurrence of liquid water on the Earth's surface would certainly be a pointer to the presence of life, but would not be sufficient evidence by itself. More important is the fact that about 20% of the atmosphere of the Earth consists of oxygen. Equally significant is the fact that the green colour that mantles large areas of the land surface is associated with complex compounds formed largely of carbon, in combination with lesser amounts of hydrogen, nitrogen, sulfur and other elements.

This mixture of chemicals is not chemically stable. Unless the carbon-rich compounds were being continuously regenerated, nearly all of this 'green mantle' would have been decomposed in the oxygen-rich atmosphere within centuries, if not decades, i.e. it would have been 'burned up'. This would have removed some of the oxygen, and the rest would have been used up in chemical weathering reactions (oxidation) involving the rocks of the Earth's surface. We must therefore conclude that some process operating on the Earth's surface continuously regenerates complex carbon-rich compounds and maintains the oxygen content of the atmosphere.

That is all very well, but we *know* there is life on Earth: the carbon-rich compounds make up the plants and the animals, and these in turn use the oxygen for respiration. So can we use the Earth as a model in our search for life on other planets in the Universe? As it happens, nobody has yet *seen* a planet outside our own solar system, although there is mounting evidence that other solar systems with planets do exist. We must begin our quest somewhat nearer home.

1.1 The search for life on Mars

Mars has been considered as a possible home of extraterrestrial life ever since 'canals' were first 'discovered' in the 19th century. They were subsequently shown not to exist, but Mars was still perceived as a possible abode for life, perhaps most famously fictionalized by H. G. Wells in his *War of the Worlds*. However, there was no way of confirming or refuting the existence of life on Mars until the advent of the space age.

In 1975, the first serious attempt was made to look for the existence of life on another planet when the US National Aeronautics and Space Administration (NASA) launched two Viking spacecraft to visit Mars. On board each of the spacecraft was a lander that was to descend through the martian atmosphere and soft-land on the planet's surface. The landers contained numerous scientific instruments and cameras that were designed to perform experiments and to answer the question: Is there life on Mars?

As the first images of the martian surface were returned to Earth in the summer of 1976 (Figure 1.2), it became clear that there were no obvious signs of life – no Martians. If there were large life-forms on Mars, then they were camera shy; but perhaps there were smaller life-forms? After all, large plants and animals have only been around on Earth for about the last 10% of its existence. Before that, the Earth's land areas were barren of vegetation and may have looked rather like the surface of Mars does today. Yet for more than three and a half billion years (see Box 1.1), i.e. for more than 75% of the Earth's existence, there have been micro-organisms on Earth. So it made sense to look for micro-organisms on Mars.

Figure 1.2
The surface of Mars as seen from one of the Viking landers.

Box 1.1 Millions and billions and units

Throughout this Course, big numbers will be expressed using the common notations of millions ($\times 10^6$) or billions ($\times 10^9$). This will apply especially to distances and ages. The latter can also be expressed in abbreviated form as Ma (megayear, million years, 10^6 yr) or Ga (gigayear, billion years, 10^9 yr). For example, 3000 million years can be written as 3 Ga or 3×10^9 yr or 3000 Ma (or even 3000×10^6 yr). Billion has sometimes been used to mean 10^{12} but this usage is no longer common.

While we use the SI system of units in this book (i.e. metres, kilograms and seconds) we shall be pragmatic and use the derived units which are fractions and multiples of these where appropriate. For example: centimetre (cm, 10^{-2} m) and kilometre (km, 10^3 m); gram (g, 10^{-3} kg) and tonne (t, 10^3 kg); year (yr, 3.154×10^7 s) and millions to billions of years (see above).

Each Viking lander was equipped with a sample arm which was designed to collect samples of the martian 'soil' (consisting of small rock fragments, sand and dust). Once withdrawn inside the lander, the sample was split between five different experiments: one to investigate the inorganic chemistry of the samples, another to look for organic molecules (see Box 1.2), and three to look for micro-organisms. However, since the only life we know anything about is here on Earth, the design of the experiments was inevitably based on Earth-bound assumptions. Each of the three microbiology experiments tested a different hypothesis about the possible metabolism of martian micro-organisms. By analogy with terrestrial life-forms, it was assumed that they must take in food and give off waste; or that they must take in gases from the atmosphere and, perhaps by using sunlight, convert them into biologically useful materials. To test the first of these ideas, samples of the martian 'soil' were mixed with a sterile source of food (a biological culture free of terrestrial bacteria) brought from Earth, in the hope that if there were any martian bugs present, they would find it palatable. Using criteria established *before* the launch of the spacecraft, two of the experiments gave positive results. One indicated that something in the martian 'soil' was chemically breaking down the organic compounds in the supply of food – the effect was similar to that seen when micro-organisms metabolize food on Earth. The other experiment exposed a martian

Box 1.2 Organic and inorganic compounds and chemistry

Some people are unclear about the distinction between organic and inorganic compounds. If the distinction puzzles you, this box should help.

The fact that carbon plays a key role in living structures accounts for the use of the term *organic* chemistry to cover the chemistry of carbon compounds. Prior to 1828, nearly all such compounds were derived directly from living or dead organisms. In that year, the German chemist, Friedrich Wöhler, made the organic compound called urea (a constituent of urine with the chemical formula NH_2CONH_2) from substances which could be obtained from non-organic (*inorganic*) matter. Thus started the detailed exploitation of organic chemistry, leading first to the development of the dye industry in the 19th century.

It is now possible to exploit the variety of behaviour found in carbon compounds to design (at the molecular level) substances that fulfil specific roles, for example, pharmaceuticals and plastics. Today, carbon-rich compounds synthesized in laboratories far outnumber those that have been isolated from natural sources. Most natural organic compounds are associated with living organisms or with their decay products after death. Some, however, are formed without the intervention of life processes (i.e. *abiogenically*) such as the organic compounds found in igneous rocks.

The distinction between organic and inorganic compounds can sometimes become blurred, but for our purposes organic compounds are combinations of carbon with variable proportions of hydrogen, nitrogen, oxygen (e.g. urea); they may contain sulfur and other elements as well. Simple molecules such as carbon monoxide (CO), carbon dioxide (CO_2) and methane (CH_4) are also associated with living systems, but of these only methane is normally classed as organic. Elemental carbon (C) itself, whether in crystalline form as diamond or as graphite, or in amorphous form as soot (see also Box 1.4), is considered to be inorganic.

sample to gases brought from Earth, and the gases appeared to combine chemically with the sample – as if there were micro-organisms taking in the gases and converting them into organic material, as we know plants do on Earth.

However, the criteria used in designing the experiments and in determining whether they were successful may have been inadequate. Back on Earth, it was found that ordinary terrestrial clays could reproduce the results of the two 'successful' Viking microbiology experiments. Clays are inorganic silicate minerals (see Section 1.2) with complex structures that can readily take up and release gases. They can also provide surfaces that are able to *catalyse* chemical reactions (a **catalyst** is a substance that speeds up or facilitates a chemical reaction but remains unchanged after the reaction). This finding also helped resolve another mystery of the Viking lander experiments: the experiment to look for organic molecules revealed no signs of such compounds in the martian 'soil'. None of the building blocks of proteins or nucleic acids that constitute the basic chemistry of life on Earth were found, not even simple molecules containing only carbon and hydrogen. Soils and many rocks on Earth (and even the Arctic and Antarctic ice-caps) contain large amounts of organic compounds derived from once-living organisms; indeed, we can find these *chemical fossils* in rocks that are at least 3.5 billion years old. The question remained, if there was life on Mars, where were its traces? The alternative conclusion, that life on Mars might be of a kind in which organic chemistry plays little or no part, seems improbable.

Carbon is abundant in the Universe, and it makes marvellously complex molecules which are ideal for life. Water is also essential: it stays liquid over a wide range of temperatures and provides a medium in which organic chemistry can take place (see Box 1.3). However, could our belief that these materials are essential for life have something to do with the fact that we are mainly made from them? Are we carbon- and water-based simply because those materials were abundant on the Earth at the time of the origin of life? Perhaps so, in which case we should seek more objective criteria in looking for life.

Box 1.3 The special properties of water

Water (H_2O) is an amazing substance. Its relative molecular mass is approximately 18. (If you are unfamiliar with the term 'relative molecular mass', see Box 2.1.) Comparison with other hydrogen-containing compounds of similar relative molecular mass suggests that water should freeze at about −100 °C and boil at −80 °C instead of at the much higher temperatures of 0 °C and 100 °C, respectively. For example, methane (CH_4), with a relative molecular mass of 16, freezes at −183 °C and boils at −162 °C; while for ammonia (NH_3, relative molecular mass 17) the temperatures are −77 °C and −33 °C, respectively. Moreover, the density of most solids is greater than that of their corresponding liquids, and the density of liquids typically decreases progressively when heated from the melting temperature. But ice is less dense than water (which is why it floats), and the maximum density of water is reached at 4 °C. The reason for these anomalous properties lies in the electrical polarity of the molecular structure of water: the uneven distribution of electrical charges on water molecules means that the molecules can arrange themselves into partly ordered structures instead of behaving as completely independent entities. The polar nature of its molecules also makes water an excellent solvent.

1.1.1 Living or non-living?

Distinguishing the living from the non-living can prove to be a problem, as the team designing the Viking microbiology experiments found. An important attribute of life that might come to mind is the amazing complexity of living things, even the simplest organism. However, complexity by itself is not enough: complex molecules exist in non-living things both on Earth and elsewhere in the Universe (as we know from meteorites and comets). Moreover, a dead mouse is, for a little while at least, almost as complex as a living one.

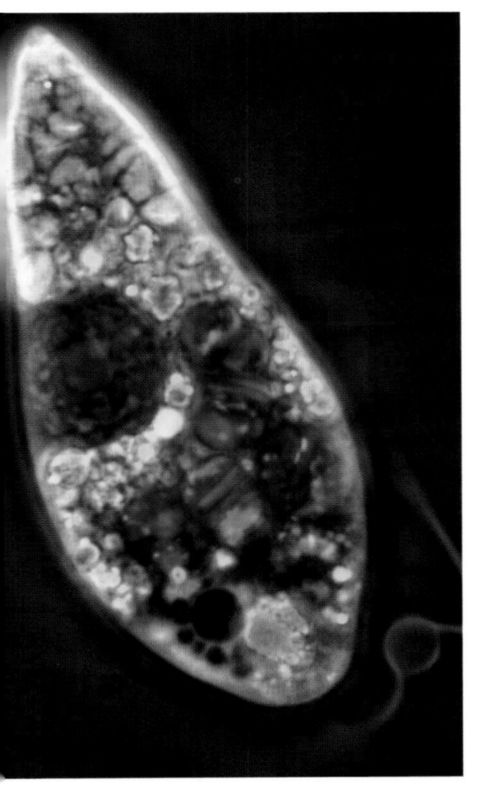

Figure 1.3
Photomicrograph of a simple single-celled organism (the photosynthetic protozoan *Euglena*). The nucleus is the brown area and the chloroplasts (the site of photosynthesis) are the green patches. The whip-like appendage is for propulsion.

Let's start by considering the essential features of a very simple unicellular organism. Figure 1.3 shows a photomicrograph of a single-celled organism that is approximately 2 μm (micrometres, 10^{-6} m) long. The cell contains millions of molecules such as fats (e.g. lipids in membranes), proteins and carbohydrates (explained further in Box 1.5), as well as RNA and DNA molecules and, of course, water. (You'll learn more about RNA and DNA and their essential roles in life processes in Chapter 4.)

This cell can do two basic things:

1. It can replicate to produce two smaller copies of itself that will then grow to the size of the original cell.
2. It can maintain itself, which involves obtaining energy and raw materials and using them to manufacture everything needed for growth, repair and replication.

The major characteristic that distinguishes *life*, therefore, is the constant renewal of a highly ordered structure. Organisms create an elegant molecular order within themselves which is adapted to carry out these life processes, and they pass on a pattern of that order to succeeding generations. The creation of this order out of chaotic surroundings is unique to life and seems to buck the general trend of the Universe whereby disorder continually increases. In fact, living creatures are like small pockets of order in an otherwise disordered Universe. The price paid, however, is the continual expenditure of energy. Hence living systems must forever take energy from their surroundings. To accomplish this, an organism must interact with its surroundings, and it is the signs of such interaction that provide evidence for the existence of life. Once the organism dies and decomposes, the energy and the chemical elements contained within it are dissipated into its surroundings.

The occurrence of such interactions between organisms and their surroundings is demonstrated by the chemically unstable situation we identified from our spacecraft at the start of the chapter. No comparable examples have been found on Mars. The density of the martian atmosphere is less than one-hundredth the density of that on Earth, and it consists almost entirely (95%) of carbon dioxide plus about 2.7% nitrogen and only traces of free oxygen (0.13%) and water vapour (0.03%) – the water appears mainly to be condensed in the form of ice at the poles. Much of the surface of Mars is rust-coloured (hence the name 'Red Planet'). This could suggest that any free oxygen which forms in the martian atmosphere is almost immediately used up in the oxidation (rusting) of iron in the surface rocks (see Box 1.4). In other words, the surface of Mars is virtually in chemical equilibrium, which would not be the case if life were present. In fact, the composition of the martian atmosphere was known *before* the Viking missions, and led James Lovelock and other scientists to predict that the biology experiments on the spacecraft would yield negative results.

- Bearing in mind the foregoing, do you think that the results of the Viking missions excluded the possibility of life ever having existed on Mars in the distant past?

- Not necessarily. There may have been organic carbon-based life-forms, but when they died they would have decayed and decomposed – in short, oxidized – and their constituent elements would have been dispersed and incorporated into inorganic compounds. Absence of *present* life thus does not preclude the possibility of past life. Indeed, as you may recall from the publicity about the possibility that there are ancient fossil micro-organisms on Mars, the results from the Viking landers were not the end of the story – a theme we develop further in Chapter 4. Some scientists even believe that some life-forms could have survived on Mars to this day, 'far from Viking's gaze'.

Box 1.4 Oxidation and reduction

As you might expect, the term **oxidation** derives from chemical reactions involving oxygen. The one you are probably most familiar with is the rusting of iron. A similar change can be brought about simply by heating metallic iron (Fe) in air, thereby producing iron oxide (Fe_2O_3). This is oxidation, because oxygen atoms have been combined with the original iron atoms to form Fe_2O_3. So a common definition of oxidation is the combination of oxygen with a substance.

Processes that involve the loss of oxygen are called **reduction** reactions. For example, the reaction of hydrogen gas (H_2) with iron oxide to produce iron metal is a reduction:

$$Fe_2O_3 + 3H_2 \longrightarrow 2Fe + 3H_2O \qquad \text{(Equation 1.1)}$$

In this reaction, the Fe_2O_3 is *reduced* to iron metal, while at the same time the hydrogen combines with oxygen to form water and is therefore being oxidized.

Iron as the metal (Fe) is in a reduced state. It can also be in a reduced state in combination with certain other elements, such as oxygen and sulfur, in the form of FeO or FeS. It is then said to be in the form of iron(II), sometimes also written as Fe(II) or Fe^{2+} (the ionic form), and it is usually called *ferrous* iron. When iron is in combination with oxygen in an *oxidized* form, as Fe_2O_3 (see above), it is said to be in the form of iron(III), sometimes also written as Fe(III) or Fe^{3+} (the ionic form), and it is usually called *ferric* iron.

Another example of oxidation is the reaction involving the conversion of carbon into carbon dioxide. For instance, although diamonds (pure carbon) may seem to be indestructible, they can be burned. If a diamond is heated to a temperature of about 900 °C, the following reaction occurs:

$$C + O_2 \longrightarrow CO_2 \qquad \text{(Equation 1.2)}$$

Unfortunately, if you wish to try to reconstitute the diamond that has been destroyed you cannot reduce the CO_2 to carbon with hydrogen and get back what you started with. The reaction may give you elemental carbon as shown in Equation 1.3, but it will most likely be in the form of an amorphous black soot-like powder rather than diamonds:

$$CO_2 + 2H_2 \longrightarrow C + 2H_2O \qquad \text{(Equation 1.3)}$$

The most fundamental process that results in reduction of CO_2 is the one that underpins nearly all life on Earth, namely **photosynthesis**. This can be most simply expressed as

$$nCO_2 + nH_2O \longrightarrow (CH_2O)_n + nO_2 \qquad \text{(Equation 1.4)}$$

where n is a whole number, usually larger than 6, and $(CH_2O)_n$ is the general formula for a carbohydrate, of which glucose ($C_6H_{12}O_6$) is the simplest (see Box 1.5).

It is important to note that the oxygen liberated in photosynthesis comes from decomposition of the *water* molecule. The hydrogen so produced then chemically *reduces* the CO_2 to form the carbohydrate molecule.

Some scientists have suggested that the earliest and most primitive organisms on Earth were nothing more than giant molecules that could direct their own replication (see Chapter 4). The reproductive process has become exceedingly complex through evolution, but its basis remains the same: one generation of an organism passes the information describing its structure and life cycle to the next. The cycle of life has been operating on Earth for at least 3.5 billion years (see Section 3.2.1). In each organism, the necessary chemical elements are assembled into the required pattern, which is passed on to succeeding generations. When an organism dies, it decays and decomposes and its constituent elements are dissipated into the surrounding environment. Thus, although every individual creature must eventually lose its own battle with chaos, the order that is inherent in life itself continues.

1.2 Life and the chemical elements

Any form of life requires large and complex molecular structures that are capable of self-replication. As mentioned previously, the organic compounds essential to life as we know it are composed mainly of carbon, hydrogen, nitrogen and oxygen (C, H, N, O); together, these constitute about 99% of living material. Table 1.1 lists the main chemical elements that make up living organisms. The major role played by water in life on Earth provides an explanation for the abundance of oxygen and hydrogen in living organisms – the human body is, in fact, 70% water by mass. However, the single most important element to life is carbon, even though it is a minor constituent of the Earth's crust (less than 0.5% by mass).

Table 1.1 The main chemical elements in terrestrial life-forms. You are *not* expected to memorize the details of this table. (Copyright © 1990 by The Benjamin/Cummings Publishing Company, Inc.)

Element	Present in all organisms	Present in some organisms
'first tier': most abundant in all organisms		
carbon (C)	✓	
hydrogen (H)	✓	
nitrogen (N)	✓	
oxygen (O)	✓	
'second tier': much less abundant but found in all organisms		
calcium (Ca)	✓	
chlorine (Cl)	✓	
magnesium (Mg)	✓	
phosphorus (P)	✓	
potassium (K)	✓	
sodium (Na)	✓	
sulfur (S)	✓	
'third tier': metals present in small amounts but essential to life		
cobalt (Co)	✓	
copper (Cu)	✓	
iron (Fe)	✓	
manganese (Mn)	✓	
zinc (Zn)	✓	

Table continues overleaf

Element	Present in all organisms	Present in some organisms
'fourth tier': found in or required by *some* organisms in trace amounts		
aluminium (Al)		✓
arsenic (As)		✓
boron (B)		✓
bromine (Br)		✓
chromium (Cr)		✓
fluorine (F)		✓
gallium (Ga)		✓
iodine (I)		✓
molybdenum (Mo)		✓
selenium (Se)		✓
silicon (Si)		✓
vanadium (V)		✓

The property of carbon that makes it such a big player in Earth and life is its chemistry. In the Periodic Table of the elements, carbon sits at the top of Group 4 with boron (B) on one side, nitrogen (N) on the other and silicon (Si) below it (Figure 1.4).

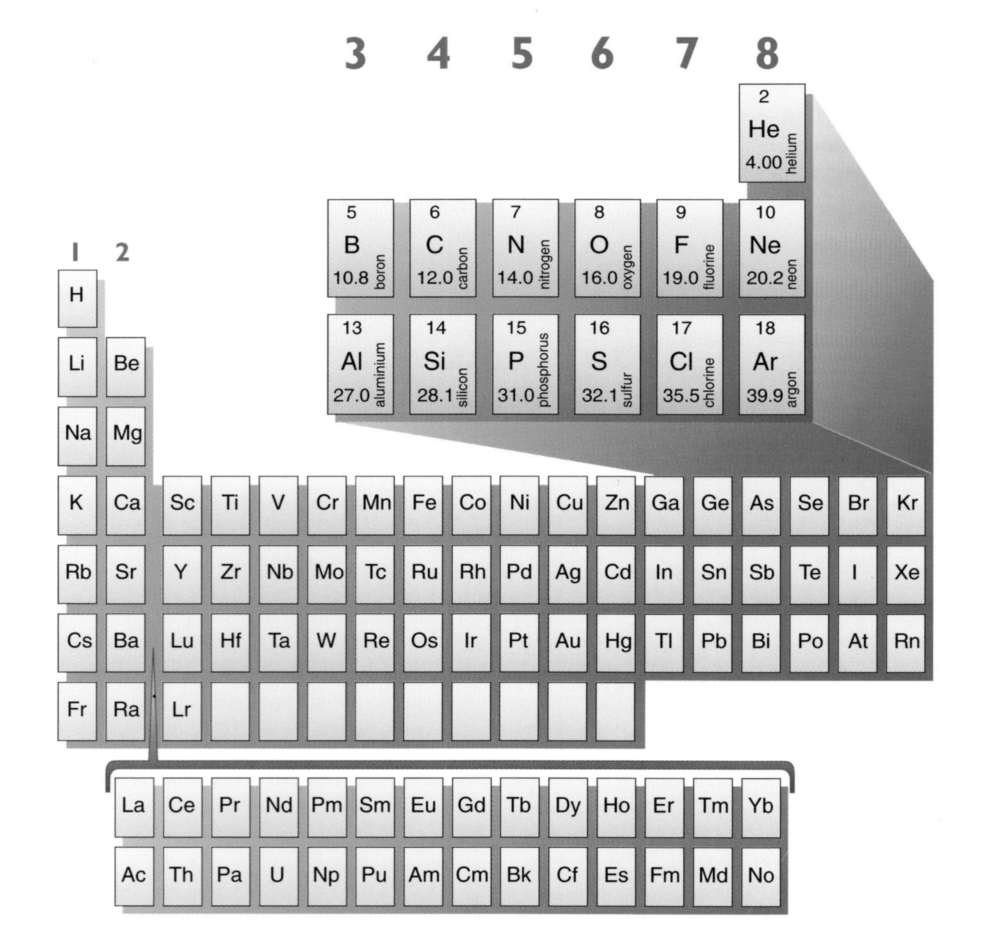

Figure 1.4
An expanded view of the upper right-hand corner of the Periodic Table, which displays the elements in order of increasing atomic number (see Box 2.1 for further details on atomic structure). A more detailed version of the Periodic Table is given in Appendix 1. The large numbers (1–8) above the columns are the Groups, e.g. carbon is in Group 4. The individual element boxes in the expanded portion contain the atomic number at the top, the relative atomic mass at the bottom, and the chemical symbol in the middle.

Carbon owes its position in the Periodic Table to the fact that there are four outer electrons in a carbon atom. This electronic structure enables carbon atoms to form up to four separate bonds with other atoms. Consider, for example, the simplest organic carbon compound, methane (CH_4). It contains one carbon atom and four hydrogen atoms (Figure 1.5a). The carbon atom forms bonds with the hydrogen atoms by sharing its four outer electrons so that the outer shell of the carbon atom acquires the full complement of eight electrons needed for a stable electron configuration – four electrons from the carbon itself and one from each of the four hydrogen atoms. (Also each hydrogen atom has, in effect, gained one electron from the carbon atom and so has the two electrons needed for it to have a stable electron configuration.) You will also notice in Figure 1.5a the tetrahedral structure of the methane molecule, and it is this three-dimensional topology of carbon that enables very complex structures to be built. Carbon readily forms similar shared-electron bonds with oxygen, nitrogen, phosphorus, and sulfur – all key elements of life.

Still more significant is carbon's ability to bond with itself. A simple example is ethanol, (ethyl alcohol) CH_3CH_2OH, in which two carbon atoms are bound together, as well as to several hydrogen atoms and an oxygen atom (Figure 1.5b). In fact, the possibilities for carbon are almost infinite: it can form complex chain-like molecules called polymers with thousands of carbon atoms joined together (Figure 1.5f), as well as complex molecules with the other elements essential to life, H, N, O (Table 1.1). Carbon atoms can even bond together to form rings thus making cyclic molecules (Figure 1.5e). Figure 1.5c shows the structure of glycine, one of the simple amino acids that are the basic building blocks for all life on Earth, as amino acids combine together to make more complex protein molecules (see Box 1.5).

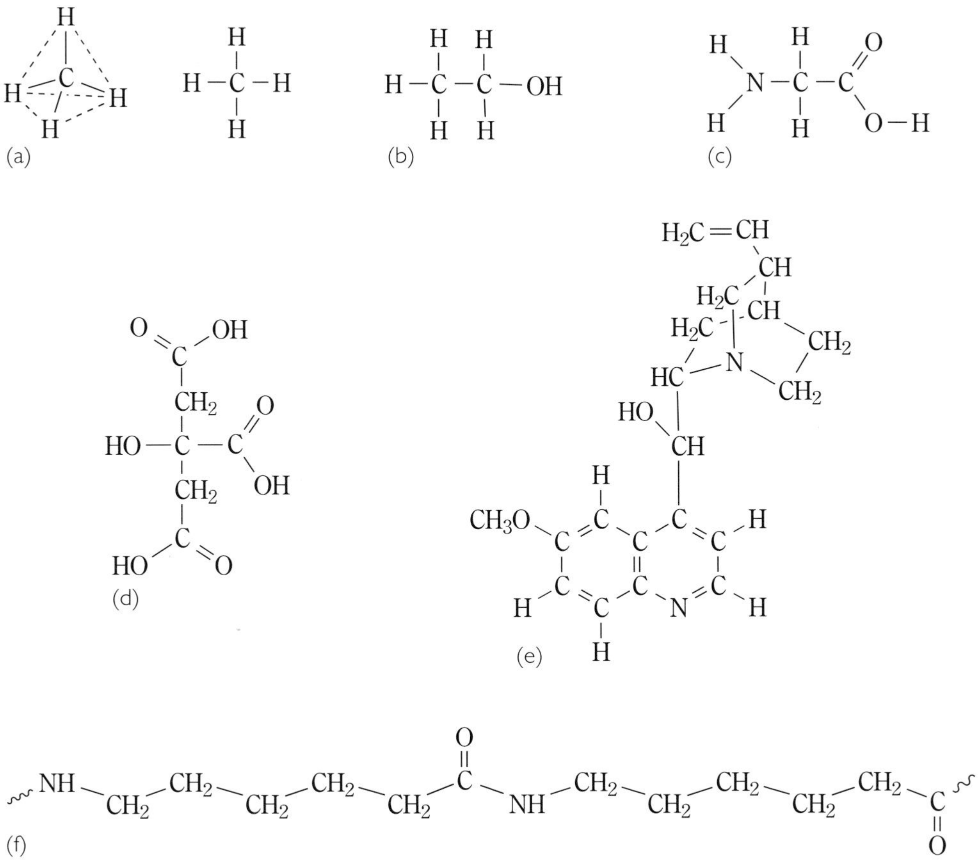

Figure 1.5 (a) The tetrahedral shape of the methane molecule (CH_4), and the same molecule represented in two-dimensions. (b) Two-dimensional representation of the structure of ethanol. (c) Two-dimensional representation of the structure of glycine, an amino acid. Two-dimensional representations of the structures of some well-known organic substances are given in the other parts of the figure: (d) citric acid, (e) quinine, (f) part of a molecule of nylon. (Three of the compounds shown, ethanol, quinine and citric acid, would be found in a gin and tonic, with a slice of lemon.) You are not expected to memorize these structures: they are included only to give you an idea of the diversity of carbon-containing compounds.

Box 1.5 The principal organic compounds necessary for life

Note: This is intended to be no more than a brief summary of information about the more important constituents of your body (and the rest of life on Earth). Don't worry if some terms are unfamiliar or if you don't understand it all. That will not hinder your comprehension of the rest of the book.

The organic compounds found in living organisms fall into four main categories: fats, carbohydrates, nucleic acids and proteins. **Fats** (and **lipids**) are the simplest, consisting of a *fatty acids* joined to a *glycerol* molecule. The molecules of **carbohydrates** are larger than those of fats, but are still relatively simple, consisting of *sugar molecules* strung together. The principal function of carbohydrates and fats in an organism is to serve as fuel – as an energy store – though plant cells are encased in rigid walls composed of a complex carbohydrate, *cellulose*. **Nucleic acids** introduce a further level of complexity. These are very large structures composed of basic units called **nucleotides** that are brought together in a great variety of proportions and sequences. Nucleic acids are the basic constituents of *genes* (the information store of nearly all organisms, see Section 4.3). The greatest variations in structure occur among the **proteins**, which include some of the largest and most complex molecules known. Proteins are built up of around 20 relatively simple amino acids (e.g. Figure 1.5c) linked together in chains hundreds to thousands of units long. The word 'protein' comes from the Greek *proteus* (meaning 'first rank') and indicates the importance of this group of molecules. Indeed, life on Earth is sometimes described as being 'protein-centred'. Proteins play a variety of essential roles in organisms: every reaction in your body is catalysed (facilitated) by a special protein catalyst (an *enzyme*). Oxygen is carried from your lungs to your tissues by a protein (*haemoglobin*) in your red blood cells. Each of the thousands of millions of cells in your body has proteins involved in every aspect of its structure and function. Organisms seem to exploit this variability in the possible structures of proteins, for no two species of living organism, animal or plant, possess exactly the same proteins – although there is a good deal of overlap.

■ How many possible combinations of amino acids are there for a protein that is 100 amino acids in length?

■ If we start by considering the combination of two amino acids, there are 20 possibilities for the first position and the same for the second. Therefore, the number of possible combinations is $20 \times 20 = 400$ for a chain of two amino acids. For a chain that is 100 amino acids long, therefore, there are 20 possibilities at the first position, 20 at the second, 20 at the third, and so on, i.e. $20 \times 20 \times 20$ ... till we reach the hundredth position. The number of possible combinations is an almost unimaginable number: $(20)^{100}$ (read as 20 to the power 100).

Consider the possible chemical basis for life without carbon. What other elements could be used? In fact, the only other element that comes close to offering us the same number of possibilities as carbon is silicon (Si), which, when combined with oxygen forms the stuff of the solid Earth, or rather of its rocky outer parts. Silicon appears in the same Group of the Periodic Table as carbon (Figure 1.4) and so a silicon atom also has four electrons in its outer shell. Like carbon, silicon is able to form complex polymers (called silicates), based on silicon–oxygen bonds, but although silicates constitute the frameworks of the minerals in rocks, they are not capable of self-replication. As you will see in Chapter 2, silicon is irreversibly locked up in minerals forming the rocky parts of planets. Moreover, its most common oxide, silica (SiO_2), is a solid at ordinary temperatures, and it is also much less soluble in water than the commonest oxide of carbon, carbon dioxide (CO_2), which is a gas at ordinary temperatures – in short, silica is less reactive under normal conditions. Silicon-based life-forms might be feasible; there might be conditions under which silicon–oxygen polymers could self-replicate, but our present knowledge of the Universe suggests that they would be confined to the realms of science fiction. So far as we can tell, the only element capable of forming self-replicating molecular structures is carbon.

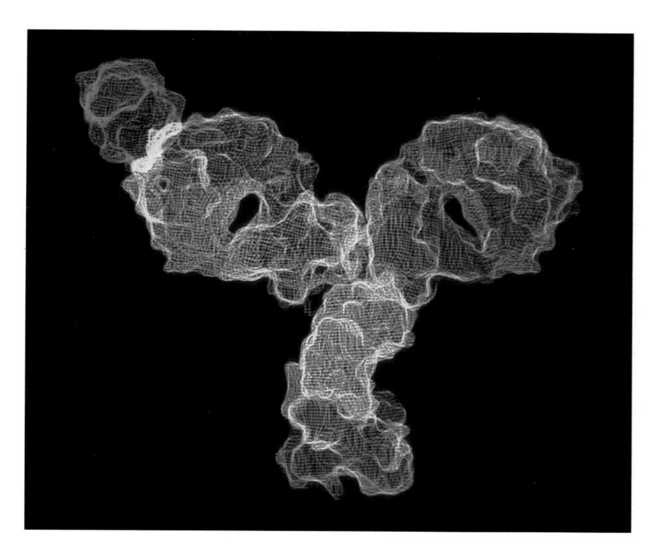

Figure 1.6 Computer graphic representation of an immunoglobin molecule (shown in green), a protein with a relative molecular mass of about 150 000, bound to an antigen – hen egg white lysozyme, another protein (shown in red).

Carbon forms a vast array of organic molecules of enormous variety and complexity (Figure 1.6 shows just one example). So, if we want to understand how organisms originated we must first explain how such complicated molecules were formed. We then need to understand how the tremendous variety of these organic molecules can be put together in the right amounts and proportions and in the right order to make living organisms. We shall return to this topic in Chapter 4, but now we consider two more fundamental questions:

First, how did carbon and other elements essential for life come to be available in the first place?

Second, why is Earth the only planet we know of where life has been able to develop and flourish – or at least to survive for so long?

The second question is no less important than the first, for without a suitably hospitable environment there can be no life. In other words, life needed a 'home' where it could develop, reproduce and evolve. So we need to consider the conditions under which chemical elements come together to form planetary bodies, as well as the conditions necessary for the formation of complex organic molecules. Both these issues are addressed in Chapter 2.

1.3 Summary of Chapter 1

1. The presence of life on Earth could be inferred from space: the coexistence of abundant free oxygen in the atmosphere and large quantities of complex organic (carbon-based) molecules at the planet's surface would suggest a condition of chemical instability. All the organic matter would be oxidized and all the oxygen would be used up, unless both were continuously regenerated. Only life is capable of such activity.
2. The essential characteristics of life are (i) the formation of complex ordered structures *that require the extraction of energy and materials* from the surrounding less-ordered environment; and (ii) the *capacity for self-replication* of these ordered structures. Upon death, the energy contained within the structures, along with the elements of which they are composed, are dissipated into the surroundings.
3. The only element known to be capable of forming ordered structures of the complexity required for life is carbon, which can combine with the other principal 'life-forming' elements (chiefly nitrogen, oxygen and hydrogen) as well as with itself, to form three-dimensional molecules of enormous intricacy and great diversity. The only other element that can form complex (polymer) molecules in this way is silicon, but its chemistry is almost exclusively based on the Si–O bond, and nearly all naturally occurring silicon-based compounds are silicate minerals that form rocks. No self-replicating silicon compounds are known. The chemistry of carbon is commonly known as organic chemistry, and all organic compounds are carbon-based, though they are not all formed by life processes.
4. The four groups of organic compounds in living organisms are *carbohydrates* and *fats*, which provide the internal 'fuel' used to supply energy; *proteins*, the largest and most complex molecular structures known, built up from the basic building blocks of amino acids; and *nucleic acids*, the basic constituents of genes (the information store).

Now try the following questions to consolidate your understanding of this chapter.

Question 1.1
If ordered structures are a requisite condition for the existence of life, why is a perfectly formed crystal not a manifestation of life – since crystals have an ordered internal structure?

Question 1.2
Carbon and silicon are vertical neighbours in the Periodic Table (Figure 1.4), both have four electrons in their outer shells, and both form polymers. Why would it be less sensible to look for silicon-based rather than carbon-based life-forms on other planets?

Chapter 2 Stars and alchemy, planets and life

The story so far:

In the beginning the Universe was created.

This has made a lot of people very angry and been widely regarded as a bad move.

Many races believe that it was created by some sort of god, though the Jatravartid people of Viltvodle VI believe that the entire Universe was in fact sneezed out of the nose of a being called the Great Green Arkleseizure. … However, the Great Green Arkleseizure Theory is not widely accepted outside Viltvodle VI and so, the Universe being the puzzling place it is, other explanations are constantly being sought.

Douglas Adams, The Restaurant at the End of the Universe *(Pan Books, 1980)*

In the beginning God created the heaven and the earth.

And the earth was without form and void; …

Genesis I, 1 and 2

Carbon is the 'stuff of life', and all life is carbon based – but carbon alone is not enough. Many other elements are essential for the existence of a living organism, however small (Table 1.1). So where did carbon and all the other elements come from in the first place? And how did they come together to form not only the planets upon which life could develop, but also the organic compounds that made life possible? Even if you allow for poetic licence and the need to tell a simple story, the biblical account of the origin of the Universe is probably nearer the mark than the suggestion made by Douglas Adams – it certainly has greater aesthetic appeal. We now understand a bit more about the cosmos than when the Bible was written, so our story is rather longer. It begins with what we know of the origins of the elements, and we start with the simplest element of all: hydrogen. You encounter the hydrogen atom every day, as a constituent of the water molecule (H_2O), which contains approximately 11% hydrogen by mass.

2.1 The airship and the hydrogen bomb

On its own, hydrogen might be considered rather dangerous. For instance, in times past, airships filled with hydrogen gas sometimes ended their days in spectacular fashion because hydrogen is very flammable (Figure 2.1).

The explosion of the *Hindenburg* was due to the hydrogen combining with oxygen to form water (i.e. the hydrogen in the airship underwent combustion – another example of oxidation, Box 1.4). This reaction can be reversed in the laboratory using a technique known as electrolysis, in which electricity is passed through water; the electric current provides energy which breaks down the water, giving the

Figure 2.1
Explosion of the *Hindenburg* in 1937 at Lakehurst, New Jersey, USA. Airships are nowadays filled with helium.

constituent gaseous elements hydrogen and oxygen. Water can also be broken down by the energy provided by ultraviolet radiation (at wavelengths of less than 200 nanometres (nm, 10^{-9} m)) from the Sun, by a process known as *photolytic dissociation*. The small amount of free oxygen in the martian atmosphere and 'soil' (Section 1.1.1) may be the result of this process. The small amount of hydrogen produced would escape to space because the gravitational field of Mars is not strong enough to retain it.

Thus, without creating or destroying matter we can convert gaseous hydrogen and oxygen into liquid water, and back again into gases. The human body is nearly 10% hydrogen by mass, mostly in water molecules, but some hydrogen is also found in complex organic molecules (Section 1.2). When we die, this 'organic' hydrogen is returned to the environment, possibly to combine with oxygen to form water, or to enter the structure of other living things. In other words, hydrogen is being continually cycled and recycled through countless chemical reactions. The same applies to all the other chemical elements. So, if the hydrogen and other elements that exist now have been cycled and recycled within and upon Earth for all of its history (in some form or another), where did they originally come from? Before we answer this, we need to bring an isotope of hydrogen, called deuterium, into the story. If you need to refresh your understanding of isotopes and of other aspects of the atomic and molecular structure of matter, read Box 2.1.

Box 2.1 Atoms, molecules, ions and isotopes

An **atom** is the smallest component (building block) of a chemical element that retains the chemical properties of that element. Atoms contain three fundamental particles: **protons** and **neutrons** (both contained within a dense nucleus), and **electrons**. Protons and neutrons have roughly the same mass, but the mass of an electron is negligible by comparison (about 2000 times smaller). An atom of the simplest element, hydrogen, has a single proton in its nucleus, which is orbited by a single electron. All atoms are electrically neutral because the positive charges on the protons balance the negative charges on the surrounding electrons. The number of protons within the nucleus defines the element and is called the **atomic number**, i.e. each element has a different atomic number (see the Periodic Table in Appendix 1). The atoms of all elements other than hydrogen also contain neutrons in their nuclei. These particles are electrically neutral. The total number of protons and neutrons in the nucleus is known as the **mass number** of the element.

Although all atoms (and molecules) are electrically neutral, they can become electrically charged if they lose or gain one or more electrons. They then become **ions**. The charge on the ion is represented by using a superscript plus or minus sign after the chemical symbol, e.g. H^+ is a positively charged hydrogen ion, Cl^- is a negatively charged chloride ion.

A **chemical compound** is a substance that consists of more than one element, with the atoms bound together in a fixed ratio by chemical bonds, e.g. water H_2O, where two atoms of hydrogen and one atom of oxygen are combined. A **molecule** is the smallest freely existing part of a substance that retains the chemical identity of the substance. Molecules do not have to contain atoms of different elements. For some elements (notably oxygen, nitrogen and hydrogen) the smallest freely existing entity is the molecule, e.g. O_2 the oxygen molecule, N_2 the nitrogen molecule, and H_2 the hydrogen molecule; whereas for others, the atoms can exist by themselves, e.g. the gases neon (Ne) and helium (He).

The chemical formula is a way of representing the constituents of a chemical compound in terms of the number and types of atoms in that compound. For compounds made up of discrete molecules, the formula represents the number and identity of atoms in the molecule, e.g. water is H_2O, methane is CH_4 and carbon dioxide is CO_2.

Isotopes

All atoms of the same element have the same number of protons in their nuclei (i.e. they have the same atomic number) but they don't necessarily have the same number of neutrons (i.e. they may have different mass numbers). These are said to be **isotopes** of the element. Isotopes of a particular element have identical chemical properties, because their properties are determined by the number of electrons around the nucleus (which is equal to the number of protons, i.e. the atomic number) and are independent of the number of neutrons in the nucleus.

If we were to look in detail at the hydrogen atoms present in water, we would find that a small proportion of them (about 0.02%) contain a neutron as well as a proton in the nucleus (Figure 2.2). This heavier isotope of hydrogen is called deuterium (and is given the symbol 'D' to distinguish it from the more common isotope, hydrogen 'H'). Hydrogen is, in fact, the only element that has named isotopes. This is potentially confusing since the element is generally called hydrogen. However, deuterium makes up only about 0.02% of naturally occurring hydrogen, so when talking about elements or chemical compositions, 'hydrogen' refers to the collective effect of both isotopes. In contrast, when isotopes or isotope ratios (see Chapter 3) are considered, 'hydrogen' refers to the isotope with mass number 1.

Isotopes are usually denoted by writing the mass number as a superscript before the chemical symbol of the element or as a number after the element name, e.g. ^2H (for deuterium or hydrogen-2) or ^{12}C (for carbon-12). A third isotope of hydrogen is called *tritium* (^3H or hydrogen-3). It has one proton and two neutrons in its nucleus, which is unstable and therefore radioactive. It is produced in nuclear reactors and nuclear explosions. Unlike tritium, the two main isotopes of hydrogen (hydrogen and deuterium) are **stable isotopes**, and so are not radioactive and will not decay with time. All other elements have more than one isotope, some of which

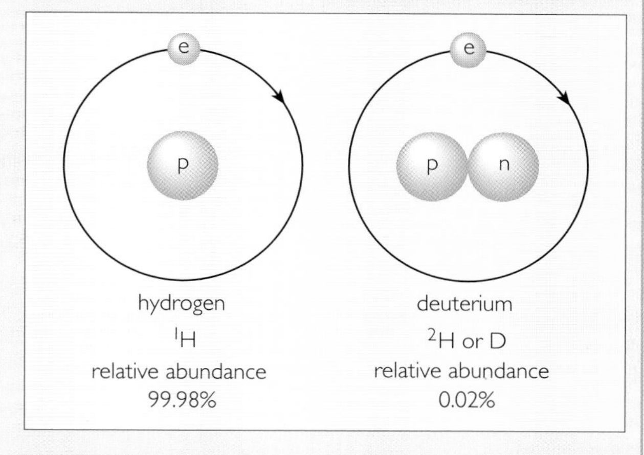

Figure 2.2
Schematic diagram of some isotopes of hydrogen. The hydrogen in naturally occurring substances on Earth is present as one of two isotopes referred to as *hydrogen* and *deuterium*. p, n, and e represent a proton, neutron and electron, respectively. (Not drawn to scale.)

may be stable, others radioactive, e.g. naturally occurring carbon contains three isotopes ^{12}C, ^{13}C, ^{14}C (carbon-12, carbon-13 and carbon-14), the first two are stable, the third is radioactive. The most abundant of these three isotopes is ^{12}C (about 99%).

Relative atomic and molecular masses

Atoms have such small masses that a convention was devised to make talking about atomic masses easy. The standard for atomic mass (the atomic mass unit) is defined as one-twelfth of the mass of an atom of ^{12}C. (1 atomic mass unit = 1.66×10^{-27} kg.)

The **relative atomic mass** of an atom is the mass of the atom expressed in atomic mass units. On this scale the relative atomic mass of carbon-12 is 12, and that of hydrogen (^{1}H) is 1.008, i.e. very close to 1 atomic mass unit (as you would expect). The relative atomic masses of the elements are included in the Periodic Table in Appendix 1.

If you look at the Periodic Table you may notice that the relative atomic mass of some of the elements such as chlorine is not close to a whole number. This is because naturally occurring chlorine contains 75% ^{35}Cl (chlorine-35) and 25% ^{37}Cl (chlorine-37). So the measured value of the relative atomic mass of chlorine is calculated from this mixture of isotopes, $[(35 \times 75)/100]+[(37 \times 25)/100] = 35.5$.

Relative molecular mass is calculated by adding together the relative atomic masses of the constituent atoms of a molecule. For example, a water molecule (H_2O) contains two atoms of hydrogen (relative atomic mass 1.01, to three significant figures) and one atom of oxygen (relative atomic mass 16.0). Therefore the relative molecular mass is calculated as $(2 \times 1.01) + 16.0 = 18.0$ (to three significant figures).

A **mole** of a substance contains the same number of particles (e.g. atoms, molecules) as the number of atoms in 12 g of carbon-12. (The mole is the SI unit of the amount of a substance and is abbreviated as mol.) This quantity is determined to be 6.02×10^{23} mol^{-1} and is known as the **Avogadro constant**. Therefore for any element or compound, an amount equal to the relative atomic or molecular mass expressed in grams will contain the same number of atoms or molecules, and that number is equal to the Avogadro constant.

■ What is the mass of one mole (1 mol) of water?

■ The relative molecular mass of water is 18.0 so the mass of 1 mol is 18.0 g.

You have probably also heard of hydrogen in the context of nuclear weapons. The first test explosion of a hydrogen bomb in 1952 released a total energy that was 1000 times greater than that from the atom bombs (Figure 2.3) dropped on Japan at the end of the Second World War.

Figure 2.3 Mushroom cloud from the testing of an atom bomb in Nevada, USA.

In the hydrogen bomb, the nuclei of two isotopic forms of the element hydrogen (hydrogen, H, and deuterium, D) become fused together by a process called *nuclear fusion*. When this happens, another element, helium, is formed and some energy is released. Hydrogen has one proton and deuterium has a proton and a neutron, so helium has two protons and one neutron (and this is known as ^{3}He, helium-3). Thus, there are fundamental differences between nuclear fusion and chemical combination. In the case of hydrogen fusion, about 10^{-12} joule (J) of energy is liberated per atomic reaction – that may seem a small amount, but it is a million times more than is liberated in the chemical combination of hydrogen and oxygen to form water. When this value is scaled up to the quantity of material in the hydrogen bomb, it translates to the release of an enormous amount of energy upon detonation. Unlike chemical reactions, during nuclear fusion mass is converted into energy. This is the basis of Einstein's famous equation, $E = mc^2$, which relates energy (E) and mass (m), with the speed of light (c).

■ Hydrogen and deuterium have mass numbers of 1 and 2, respectively, and helium has a mass number of 3 (i.e. $^{1}H + ^{2}H = ^{3}He$). Can you suggest how mass can have been converted into energy during this nuclear fusion reaction, according to Einstein's equation?

■ It is in the nature of nuclear fusion reactions that the whole really is 'less than the sum of its parts'. One proton (hydrogen) fused with another proton plus a neutron (deuterium) do indeed make a nucleus with two protons and a neutron (helium-3), but a small amount of mass is lost in the transaction, i.e. it has been converted into energy according to Einstein's equation.

The 'missing mass' is actually infinitesimal, but the speed of light is so enormous (approximately 3×10^8 m s^{-1}) and c^2 is even bigger (9×10^{16} m^2 s^{-2}) that the amount of energy liberated by the loss of even a tiny amount of mass makes nuclear fusion a truly awesome source of energy.

Question 2.1
Tiny amounts of mass really *are* involved in nuclear fusion reactions. Given that the energy liberated per individual H–D fusion is 10^{-12} J, use the equation $E = mc^2$ to work out how much mass is converted into energy per individual H–D fusion. The SI unit of energy is the joule (J), which is equivalent to the basic SI units of kg m^2 s^{-2}.

The fact that the atoms formed during this fusion reaction are composed of two protons instead of one shows that another element with different chemical properties has been formed. Helium is different from hydrogen: although it is also a light gas, it is inert (i.e. it doesn't react readily) and forms no natural compounds. Helium has two isotopes, ^{3}He (helium-3) and ^{4}He (helium-4); the latter can also be formed by nuclear fusion, as you will see shortly.

- How many protons and neutrons are there in ^{4}He?
- There must be two protons, because it is an isotope of helium. As it has a mass number of 4, there must also be two neutrons.

The H–D fusion reaction described above also occurs naturally in stars and for some of them it is their main source of energy. There it is part of the process known as **hydrogen fusion**. (Amongst physicists, hydrogen fusion is sometimes called hydrogen burning. We will not use this term because of the potential for confusion with the term 'burning' as applied to oxidation, which is a *chemical* reaction.) In the same way that the exploding hydrogen bomb releases a massive amount of energy, so nuclear reactions in stars produce heat and light – which is why stars shine. Hydrogen fusion, with the consequent production of helium, is just one example of the nuclear reactions taking place in stars. We will consider some of the others below, but the central point to remember is: *All stars start predominantly as the simplest of all elements, hydrogen and helium, and evolve to produce heavier elements.* Stars are the embodiment of the alchemist's dreams, manufacturing – amongst other things – gold, silver and platinum. If this seems a long way from our quest for the origin of life on Earth, bear in mind that there would be no alchemists (or anyone else) to dream unless the stars also made carbon – and you will soon find out how that is done.

Question 2.2
What would have been the consequences of the *Hindenburg* disaster if the airship had been filled with deuterium instead of hydrogen?

2.1.1 Our Sun – a helium factory

The star in our own solar system, the Sun, provides the light and heat that are essential for life on Earth to be sustained. The Sun is a fairly typical star, a massive ball of material, mostly hydrogen with a few per cent of helium, undergoing hydrogen fusion (as are many other billions of stars in our galaxy and the Universe). Pressures and temperatures in the interior of the Sun are so great (10^{16} Pa and 10^7 K, respectively, see Box 2.2) that all the constituent molecules have been broken down into atoms, e.g. all the hydrogen exists as H atoms and not H_2 molecules. In detail, the atoms have lost their electrons and as a result, all the elements are present in the form of ions. This mixture of ions and electrons all squashed together results in a state of matter known as a *plasma*. A plasma is sometimes referred to as the 'fourth state of matter', to distinguish it from the other three states with which we are more familiar (see Box 2.5): solid, liquid and vapour (gas).

Clearly, the type of plasma present within the body of the Sun is something we do not normally come across on Earth. However, we can find examples of plasma in our everyday experience. For example, the light emitted from neon signs is the result of a plasma. In this instance, the plasma is formed by the electrical excitation of neon at low pressures, producing a mixture of ions and electrons, and the *light* comes from the electrical excitation of the ions in the plasma so produced. The difference between such plasmas and that in the Sun is that in the latter, the constituent ions are under such high pressure that individual nuclei can fuse together.

Figure 2.4 is a schematic diagram showing how hydrogen fusion begins with two hydrogen nuclei joining to form deuterium, thereby releasing energy. This deuterium nucleus subsequently fuses with another hydrogen nucleus to form ^{3}He (helium-3). Ultimately, two ^{3}He nuclei fuse together to form ^{4}He (helium-4), the other isotope of helium, and at the same time two more hydrogen nuclei are expelled. The expelled hydrogen nuclei are then able to take part in further hydrogen fusion reactions, releasing yet more energy.

The Sun may be a small star by cosmic standards but it is still huge. It has a mass of 2×10^{30} kg and contains plenty of hydrogen.

- Can hydrogen fusion in the Sun continue indefinitely?

- No: eventually the hydrogen in the Sun will all have been converted into helium.

This is happening to stars throughout the Universe all the time – they are running out of hydrogen. What happens to them next is very much dependent upon the mass of the individual star. If it is relatively small, nuclear reactions cease altogether and the star dies, i.e. it cools down and contracts under gravity. On the other hand, if the star is large enough and has sufficiently high pressures and temperatures within its core, another form of nuclear fusion takes place: **helium fusion**. Our Sun is just large enough for helium fusion to occur, but this will not happen for perhaps another few billion years.

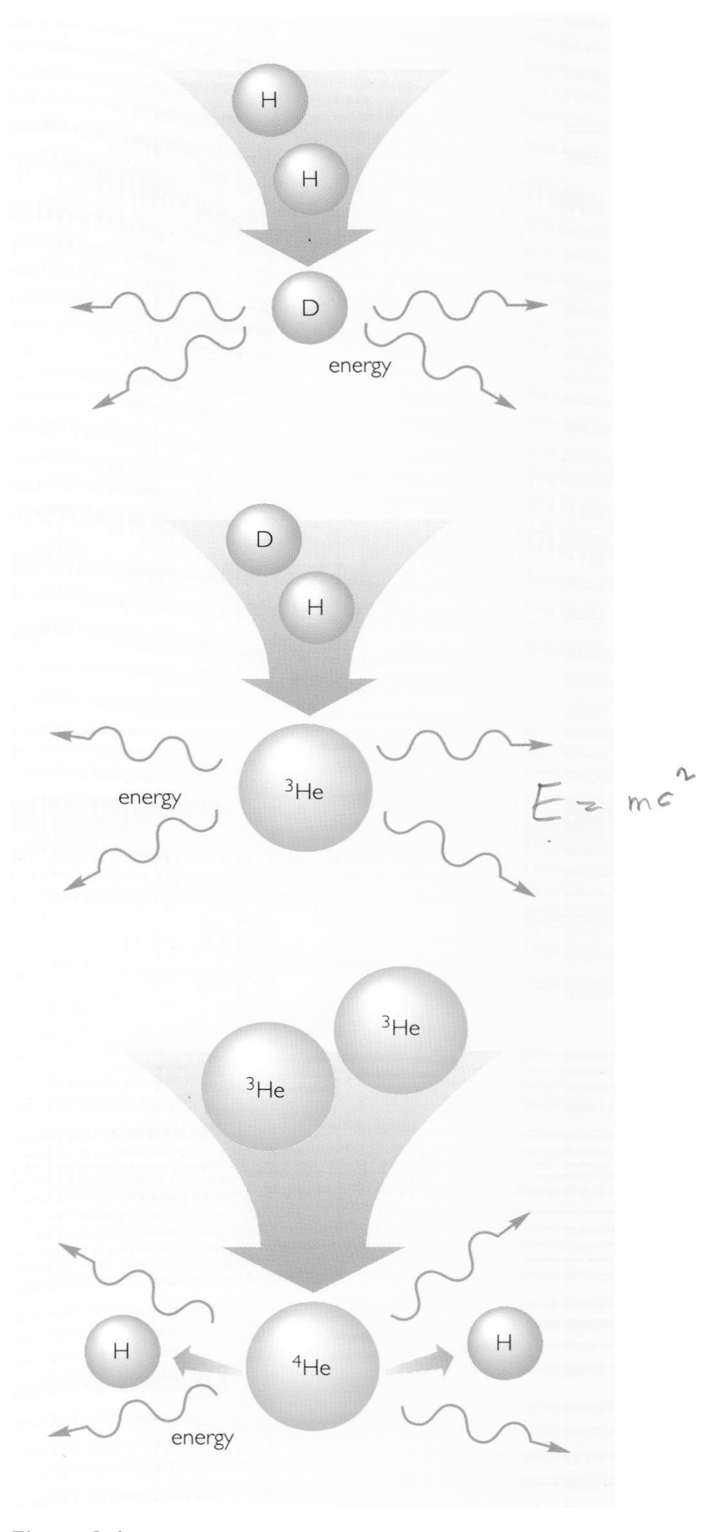

Figure 2.4
Schematic diagram of the hydrogen fusion reactions that take place in stars. H = hydrogen, D = deuterium, He = helium. (Not drawn to scale.)

Box 2.2 Pressure and temperature

Pressure is measured in newtons per square metre, $N m^{-2}$, where the *newton* is a unit of force, expressed in basic SI units as $kg m s^{-2}$. The basic unit of pressure thus becomes $kg m^{-1} s^{-2}$ (i.e. $kg m s^{-2} m^{-2}$), which is also known as the pascal (Pa). Normal atmospheric pressure at the Earth's surface is 10^5 Pa. So the pressure at the centre of the Sun, 10^{16} Pa, is eleven orders of magnitude greater than that at the Earth's surface. For comparison, the pressure at the Earth's centre is about six orders of magnitude greater than that at the Earth's surface, and in the deepest part of the ocean it is about three orders of magnitude greater.

Temperatures are generally expressed in degrees Celsius (centigrade), °C. At sea-level, pure water freezes at 0 °C and boils at 100 °C. Absolute zero, the lowest possible temperature, where all atomic and molecular motion ceases, is −273·15 °C, and this is used as the starting point (i.e. zero) for the **Kelvin temperature scale**. (It is worth noting that this temperature has never been achieved.) The unit of temperature on this scale is the kelvin, K, *not* °K. Intervals on the Kelvin scale are exactly the same as on the Celsius scale (i.e. 1°C = 1 K), so 0 °C is equivalent to 273·15 K and 100 °C is equivalent to 373·15 K. A comfortable (warm) temperature at the Earth's surface is 300 K or 27 °C – though some might think that a bit hot. The temperature at the centre of the Sun, 10^7 K, is five orders of magnitude greater than this and between three and four orders of magnitude greater than that at the centre of the Earth.

2.2 Making the heavier elements

During helium fusion, three ^{4}He nuclei fuse together to form ^{12}C (carbon-12), the principal isotope of carbon. Subsequently, ^{4}He nuclei can fuse with ^{12}C nuclei to form the principal isotope of oxygen, ^{16}O (oxygen-16). This is the route to heavier elements (see Section 2.2.2), but first we look at a nuclear fusion process that is very important in the context of life. It is a striking fact that the major life-forming elements, carbon, nitrogen, oxygen and hydrogen, which are close neighbours in the Periodic Table (Figure 1.4), are also inextricably linked in the very places where they are made.

2.2.1 The stuff of life

Helium fusion in a star can actually begin before it runs out of hydrogen (Section 2.1.1). The helium fusion is confined to the core region of the star, while the surrounding material is still composed predominantly of hydrogen. It is possible for the products of helium fusion in the core to mix with material from the outer layers, which means for example that ^{12}C can mix with hydrogen. This opens up a whole range of possibilities for the synthesis of the elements central to life, in a process known as the **CNO cycle** (Figure 2.5). First, hydrogen undergoes nuclear fusion with ^{12}C to form the other stable isotope of carbon, ^{13}C. Subsequent fusions with hydrogen produce the isotopes of nitrogen (^{14}N and ^{15}N), and of oxygen (^{16}O, ^{17}O and ^{18}O).

The CNO cycle can be broken at any stage, and elements of progressively higher mass produced (see Section 2.2.2). Within the cycle itself, however, the addition of hydrogen to an element can result in the simultaneous expulsion of a helium nucleus. In this case, there is a net loss of subatomic particles and an element of lower mass is formed (e.g. ^{15}N from ^{18}O, or ^{12}C from ^{15}N). Thus, *within* the cycle there is no net gain or loss in the total number of carbon, nitrogen and oxygen (CNO) nuclei (i.e. the *total* number of nuclei of C + N + O remains the same throughout) – they merely transform from one element to another, with the CNO nuclei facilitating the conversion of hydrogen to helium.

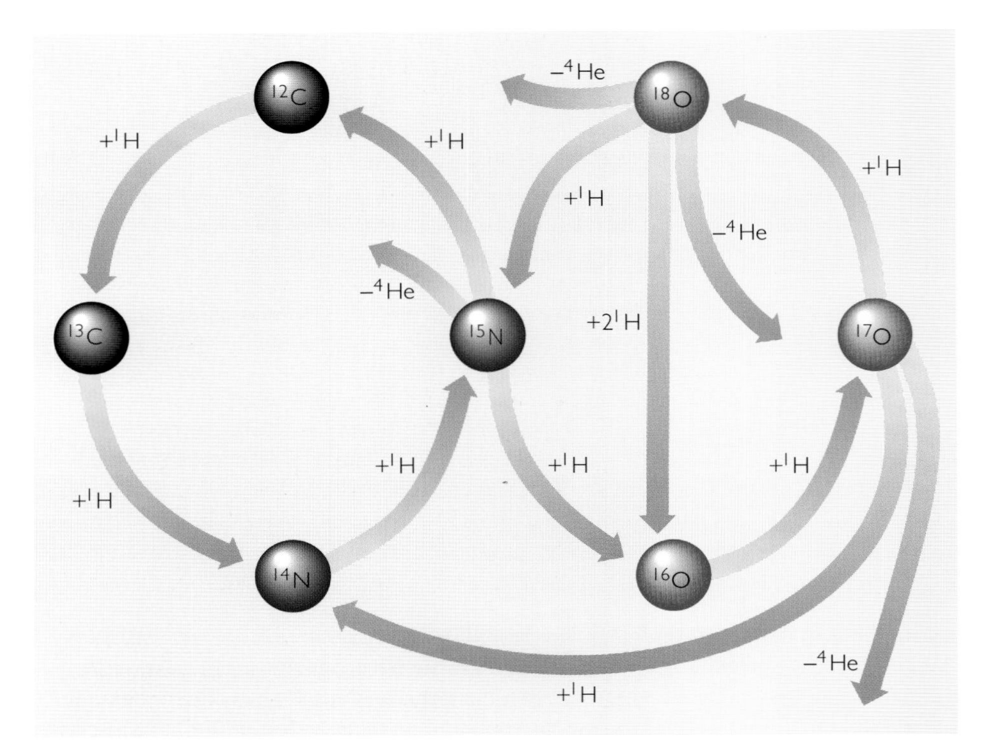

Figure 2.5
Schematic diagram of the CNO cycle. The addition of hydrogen ($+^1H$) to an element can result in the expulsion of a helium nucleus ($-^4He$), so forming an element of lower mass.

2.2.2 Other heavier elements

If the ^{12}C nucleus produced by helium fusion then fuses with another ^4He nucleus, instead of entering the CNO cycle, the principal isotope of oxygen (^{16}O) is formed. If this in turn fuses with another ^4He nucleus, ^{20}Ne (neon-20) is produced as shown in Figure 2.6. Note that the ^{16}O formed here is by direct fusion of a helium nucleus with ^{12}C and *not* via the CNO cycle.

In stars that are large enough (and thus sufficiently hot), carbon nuclei can fuse together. In this case we have carbon fusion, which produces elements such as sodium and magnesium. Additionally, elements such as sulfur, phosphorus and silicon may be produced by oxygen fusion. Fusion of elements such as carbon and oxygen often occurs in an explosive manner, giving rise to events known as novae and supernovae. A **nova** is an explosive event that takes place in a star which is usually one part of a binary system (two physically close stars), and results in the rapid ejection of a small part of its mass to the other star in the binary system. Such events are visible to astronomers because the process is accompanied by an increase in luminosity during the mass exchange. A **supernova** is much more spectacular: a whole star contracts as it runs out of nuclear fuel, increases its speed of rotation and eventually explodes, sending a large proportion of its mass out into space. Novae and supernovae are responsible for the production of many other elements and isotopes.

Nuclear fusion can produce all elements with masses between that of ^{12}C (carbon-12) and ^{56}Fe (iron-56) (see the Periodic Table in Appendix 1), although some researchers consider that the heaviest element formed in this way is ^{58}Ni (nickel-58). However, there are another 60 or so elements which are heavier than iron (or nickel). These are formed predominantly by a process known as *neutron capture*, in which neutrons are continually added to a particular nucleus, making it progressively

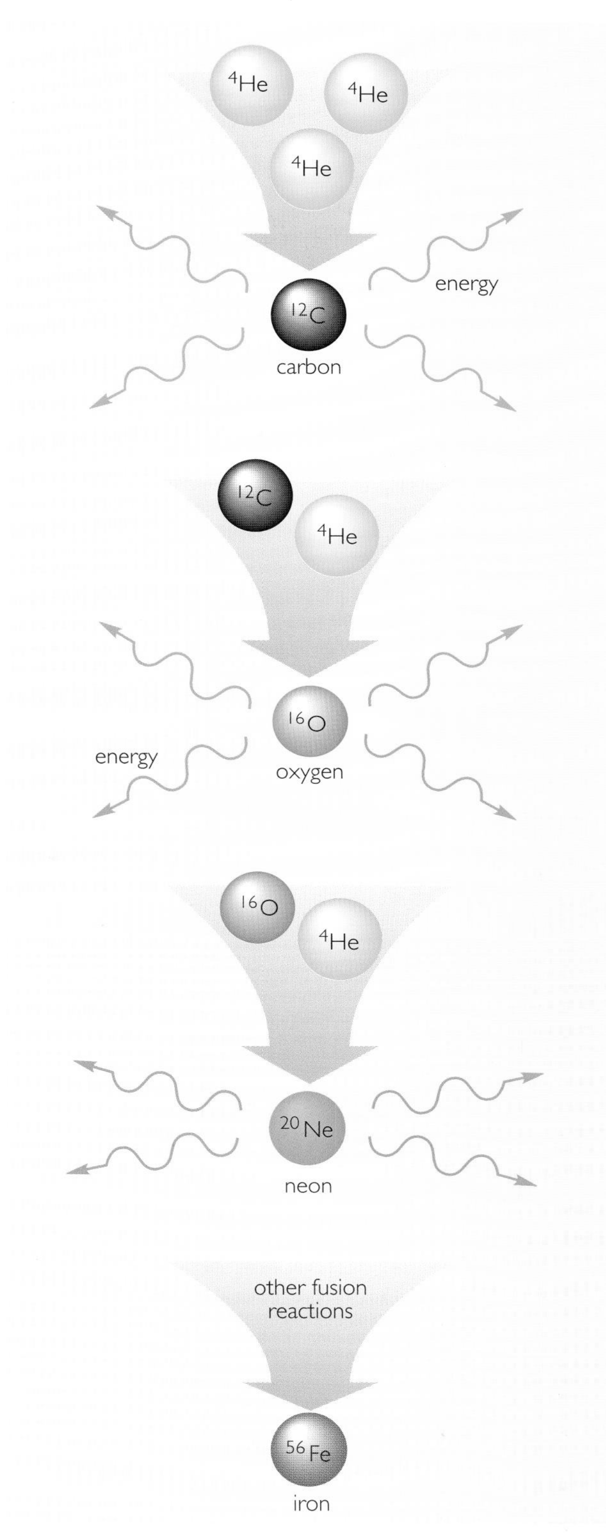

Figure 2.6
Schematic diagram of the helium fusion reactions that take place in stars, resulting in the formation of ^{12}C, ^{16}O and ^{20}Ne. Other nuclear fusion reactions produce elements with masses up to that of ^{56}Fe.

heavier, until it eventually becomes unstable and undergoes radioactive decay. If this decay involves expulsion of a high-energy electron *from the nucleus*, then a neutron is transformed into a proton and a new element is created, as the atomic number increases by one. (For present purposes, we can think of a neutron as consisting of a proton and an electron, so if you can lose the electron you leave a proton.) The continuation of neutron capture eventually produces elements all the way up to uranium, ^{238}U (uranium-238), the heaviest element that exists naturally on Earth.

Question 2.3

It is possible to continue adding neutrons to uranium and ultimately to produce heavier elements. Indeed, this can be performed on Earth in nuclear research laboratories, and almost certainly takes place in stars. Why, therefore, do you think these elements are not found naturally on Earth?

A favourable site for neutron capture is within a supernova, where the reactions take place rapidly, and the newly created materials are ejected violently into space. Neutron capture can also occur within some larger stars which have a layered structure with, for instance, concentrations of helium and carbon in the core where helium fusion is taking place, while hydrogen fusion occurs in a layer around this. In such cases, the reactions take place slowly, and the products are expelled at a more or less steady rate as streams of charged particles called *stellar winds.* In fact, all stars, including our own Sun, produce stellar winds (see Section 2.3.1).

Question 2.4

(a) How and where are chemical elements with masses less than that of iron formed?

(b) How are the elements that are heavier than iron formed?

(c) How is carbon *principally* synthesized within stars?

2.2.3 Hydrogen – the *sine qua non*

We have seen how hydrogen fuels the stars and how hydrogen is converted by nuclear fusion reactions into helium, which is then transformed to carbon, oxygen, neon, magnesium, and so on up to iron (or nickel). And we have seen how the heavier elements are formed by neutron capture processes. But where did all the hydrogen come from in the first place?

This is where we have to appeal to theoretical physicists, philosophers or theologians, depending upon our outlook. Scientifically, the consensus explanation involves formation in the so-called Big Bang. The Big Bang can be loosely defined as an instant in time when all the matter of the Universe, which was previously compressed into a single point, began to expand rapidly. Cosmologists are not yet agreed about whether this 'singularity' came into being through the gravitational collapse of some previous Universe. Although it is difficult to comprehend, the hydrogen nuclei are all considered to have formed within one second of the initiation of the Big Bang. Within the first million years, some deuterium and helium were also produced. From then on, galaxies and stars were formed and the creation of all the other elements began, involving the birth and death of many generations of stars in a truly cosmic cycle.

2.3 The cosmic cycle

In the preceding section, we saw how the chemical elements are formed either within stars (by processes of nuclear fusion, CNO cycling, neutron capture, and novae outbursts) or in explosive processes that effectively destroy stars (such as supernovae).

- How many of the elements present on the Earth were originally formed in the Sun?

- You might suspect that the answer is hydrogen and helium. But in fact the answer is none. You will see why in the next section.

2.3.1 Secondhand elements

It is true that the constituents of the Earth were formed in stars that were, in some cases, *like* the Sun. However, all of the stellar processing and nuclear reactions necessary to produce the elements on Earth took place *before* the Sun existed. Even though the Sun has been synthesizing elements during its lifetime (mainly helium, but also other light elements such as lithium, beryllium and boron), none of these have reached the Earth: almost all of the newly formed products have remained within the body of the Sun. Nevertheless, relatively small amounts of material *have* been lost from the Sun, expelled into space in all directions as charged atomic particles, forming the **solar wind**. In fact, these particles never reach the surface of the Earth because our planet has a strong magnetic field which deflects them, although some can become trapped at very high altitudes of several hundred kilometres. Here, the atmosphere is very thin and some of the gases are ionized as a result of excitation by solar radiation. When solar wind particles cause further ionization (a process that takes place predominantly at high latitudes), light energy is produced and we see the multicoloured lights of the polar aurorae (like the plasma in a neon tube excited by electrical discharges, Section 2.1.1). In contrast, the Moon, having no magnetic field, is constantly bombarded by the solar wind, which is estimated to carry some 10^{33} particles to the Moon's surface every year. Indeed, rocks from the Moon (brought to Earth by space missions) have been studied to determine the evolutionary history of the Sun.

Take a look at Figure 2.7, which gives you a basic outline of the region of space within which we live. As you look at it, bear in mind the following points:

1. The constituent materials of the planets and asteroids were produced by astronomical processes that pre-date the Solar System.
2. The Sun is *also* composed of materials that have come from previous generations of stars. In other words, the Sun contains elements such as carbon, even though helium fusion has not yet started there. The carbon in the Sun was formed by helium fusion in stars that no longer exist.

The names of the planets are probably familiar to you, but you may be less familiar with asteroids, in which case refer to Box 2.3.

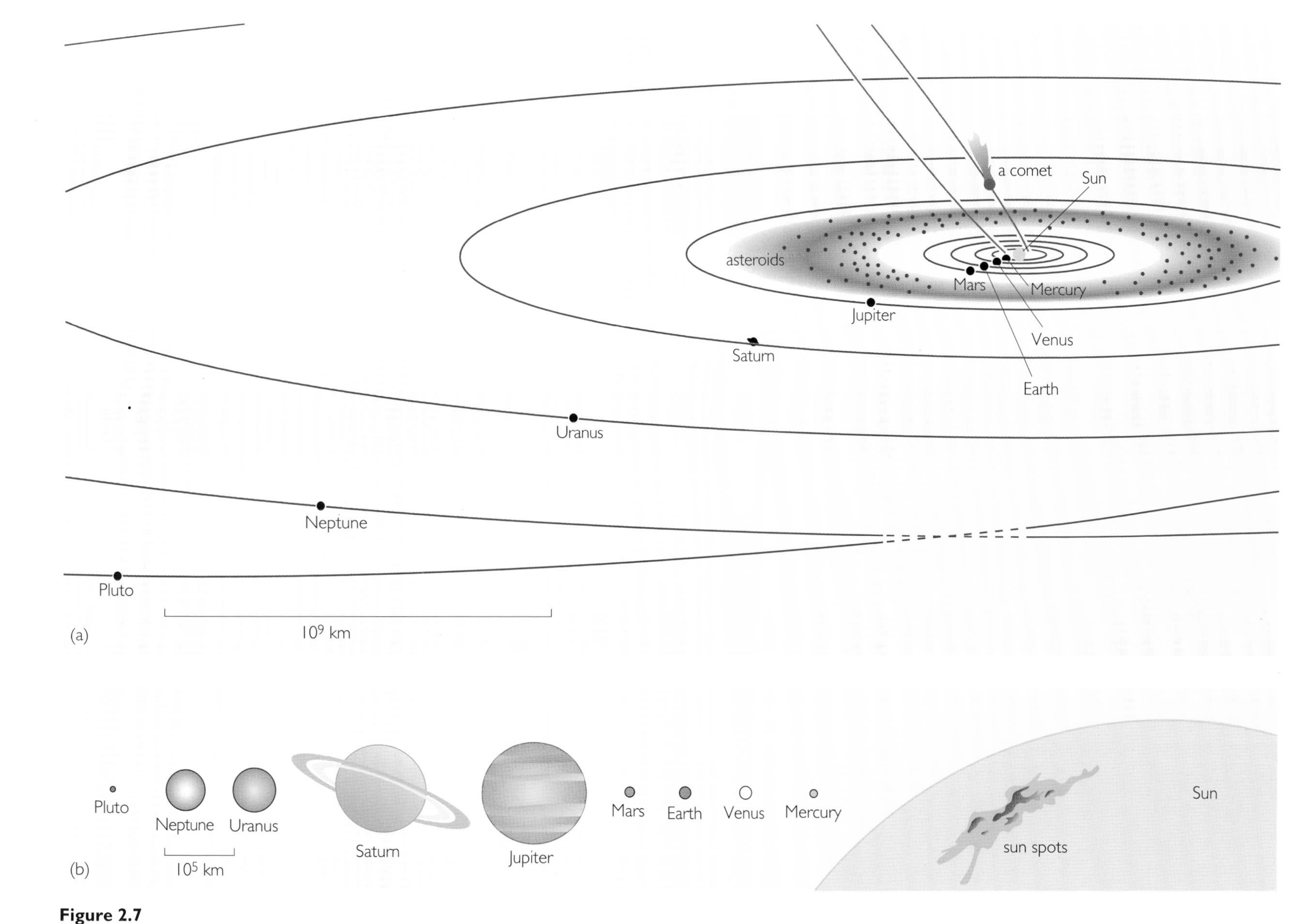

Figure 2.7
Basic layout of the Solar System.

Box 2.3 Asteroids and meteorites

The **asteroid belt** is a collection of minor bodies orbiting the Sun between Mars and Jupiter (Figure 2.7). **Asteroids** are typically a few kilometres in size, ranging up to 457 km. Their small sizes indicate that they cooled and crystallized early in the history of the Solar System, and they have not subsequently been affected by surface or internal processes. Along with comets (see Section 2.4.5), they are the 'fossilized' remains of Solar System formation. The asteroids were once thought to be the debris of a hypothetical planet between the orbits of Mars and Jupiter that broke up at some time in the distant past. However, it now seems clear that they consist of solid material that failed to aggregate into a planetary body. The most likely cause of an asteroid leaving its orbit in the asteroid belt and striking the Earth is a perturbation of its orbit by the gravitational field of one of the larger planets. If, say, Jupiter and Saturn, which travel round their orbits with different periods, happen to be in the same part of the Solar System at the same time, perhaps even 'lined up' with the Sun, their combined gravitational fields could be sufficient to pull small bodies out of the asteroid belt and send them into quite different orbits. Some, at least, would be placed in orbits away from Earth – indeed, it is possible that one or more moons of both the giant planets (Jupiter and Saturn) are asteroids captured in this way. Some, however, end up in orbits that take them towards the inner planets, including Earth.

Gravitational perturbation of their orbits can also cause asteroids occasionally to collide with one another, and such collisions are thought to have occurred more frequently in the past. When these collisions occur, showers of fragments called *meteoroids*, ranging in size from dust to large boulders, are scattered into space. A tiny fraction of these meteoroids reaches the Earth, but most of them are small and vaporize in the atmosphere as a consequence of the heat generated by friction – a process sometimes called *ablation*. Their incandescent trails are the 'shooting stars' seen in the night sky (though not all shooting stars are from the asteroid belt, see Section 2.4.5). A tiny fraction, however, survive to reach the Earth's surface, and these are called **meteorites**. Of these only a small number are found, as most meteorites are a few centimetres to a few metres in size (although much larger bodies than this have hit the Earth, with devastating consequences, see Chapter 3). They often have an outer skin that shows signs of melting. From analysis of meteorites, we know they are unlikely to have a single source, for they have a great range of chemical and mineralogical compositions: from almost pure metal (an iron–nickel alloy), through mixtures of metals with silicate minerals, to mixtures of different kinds of silicate and sulfide minerals in a wide range of proportions and textures. As you will see in Section 2.3.2, some of them also contain complex organic compounds. Meteorites have proved invaluable for documenting our origins: they enable us to date the formation of the Solar System, and they provide insights into processes that took place in the past.

When the Sun eventually runs out of nuclear fuel, it will begin to expand, forming an object known as a *red giant*, and the flux of particles in the solar wind will increase. The planets of the Solar System will be consumed and/or destroyed by the expanding red giant, and decomposed into their constituent elements, before being added back to space (Figure 2.8). There, the debris of our solar system will mix with the remains of other astronomical bodies (perhaps even with the remains of previous planetary systems that may once have supported life). Out of this cosmic cauldron, new stars and planetary systems will be born, and life may develop and evolve upon some of the newly formed planets. In short, our solar system is part of a grand **cosmic cycle** in which the physical appearance and chemical composition of the components are constantly changing.

We do not know how long the cosmic cycle will continue to operate, but we can be fairly confident that there are at least several billion years left. However, we do have some notion of the *age* of the Universe and, in particular, of our own solar system, i.e. how long they have been in existence.

The age of the Solar System, and hence of the Earth, has been reliably determined from meteorites to be about 4.6×10^9 years (i.e. about 4.6 billion years). The age of the Universe, on the other hand, is open to considerable debate because of the method used to determine it. This involves observations of very distant galaxies; these observations depend on the nature of the equipment used, and are subject to

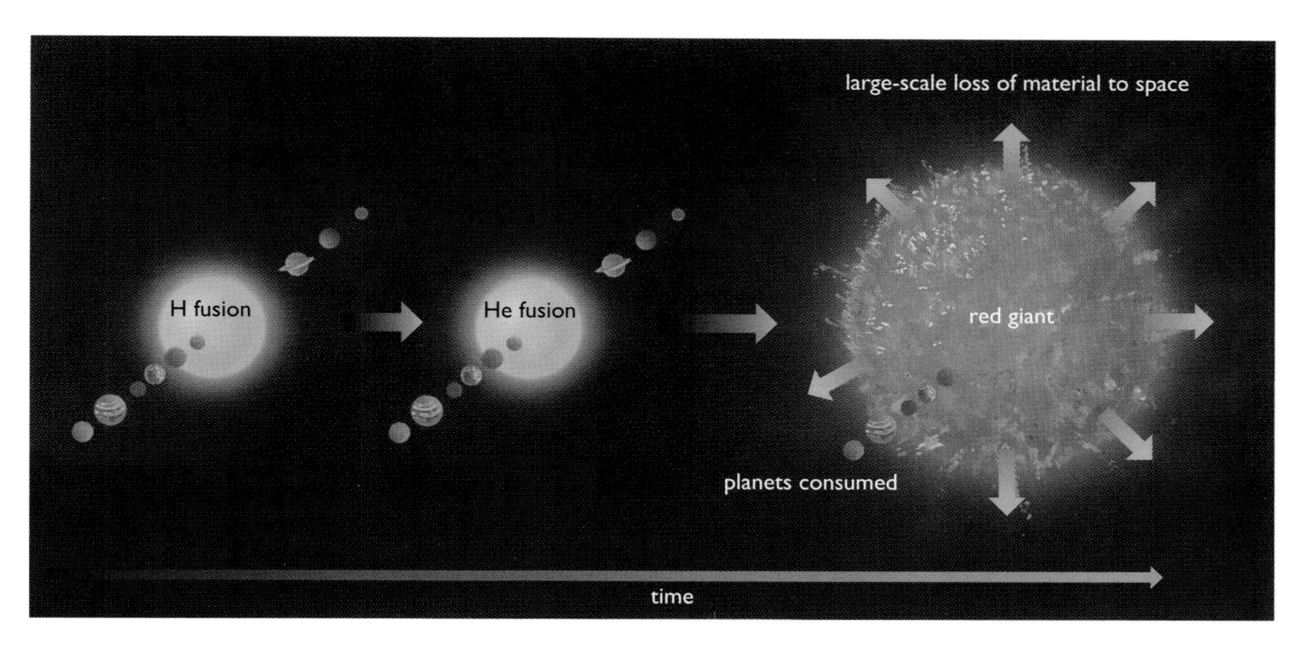

Figure 2.8
A look into the future. How the Sun will evolve and ultimately die. The present-day Sun is illustrated on the left, and the time-scale arrow represents several billion years. This diagram is not to scale – the planets are shown schematically and the size of the red giant is such that it would extend a considerable distance towards the Earth.

different assumptions and interpretations and probably to one or two prejudices as well. However, the consensus view is that the Universe is about 15×10^9 years old, with an uncertainty of perhaps $\pm 5 \times 10^9$ years.

- Can you place upper and lower limits on the age of the carbon in our bodies?
- It must be older than the age of the Solar System ($>4.6 \times 10^9$ years) but less than that of the Universe ($<15 \times 10^9$ years).

2.3.2 The cosmic abundance of the elements

Let us proceed with the generally accepted view that around 15×10^9 years ago the presently observable Universe was created out of the Big Bang. Hydrogen was the dominant element made at this time, but some deuterium and helium were also formed. As the Universe started to expand and cool, primordial matter began to form locally into concentrations which ultimately spawned galaxies (Figure 2.9), within which the very first stars were born. In stars large enough to undergo helium fusion, elements such as carbon and oxygen were made; these elements could then take part in CNO cycles as well as making still heavier elements. Novae and supernovae began to eject all kinds of heavy nuclei, while neutron capture in some stars also made heavier elements, which were gradually released into space by stellar winds, along with the products of CNO cycling. Thus, from an environment once dominated by hydrogen and helium, progressively more and more chemical elements were produced in stellar 'factories', novae and supernovae.

The vast size of the Galaxy, let alone the Universe (see Box 2.4), means that trying to establish an overall chemical composition is a difficult task. But we can at least make a start within the Solar System that we inhabit.

- What might we use to begin making an estimate of the abundance of elements in the Solar System?
- We can use meteorites that fall on the Earth's surface, at least some of which might be representative samples of 'average' Solar System material.

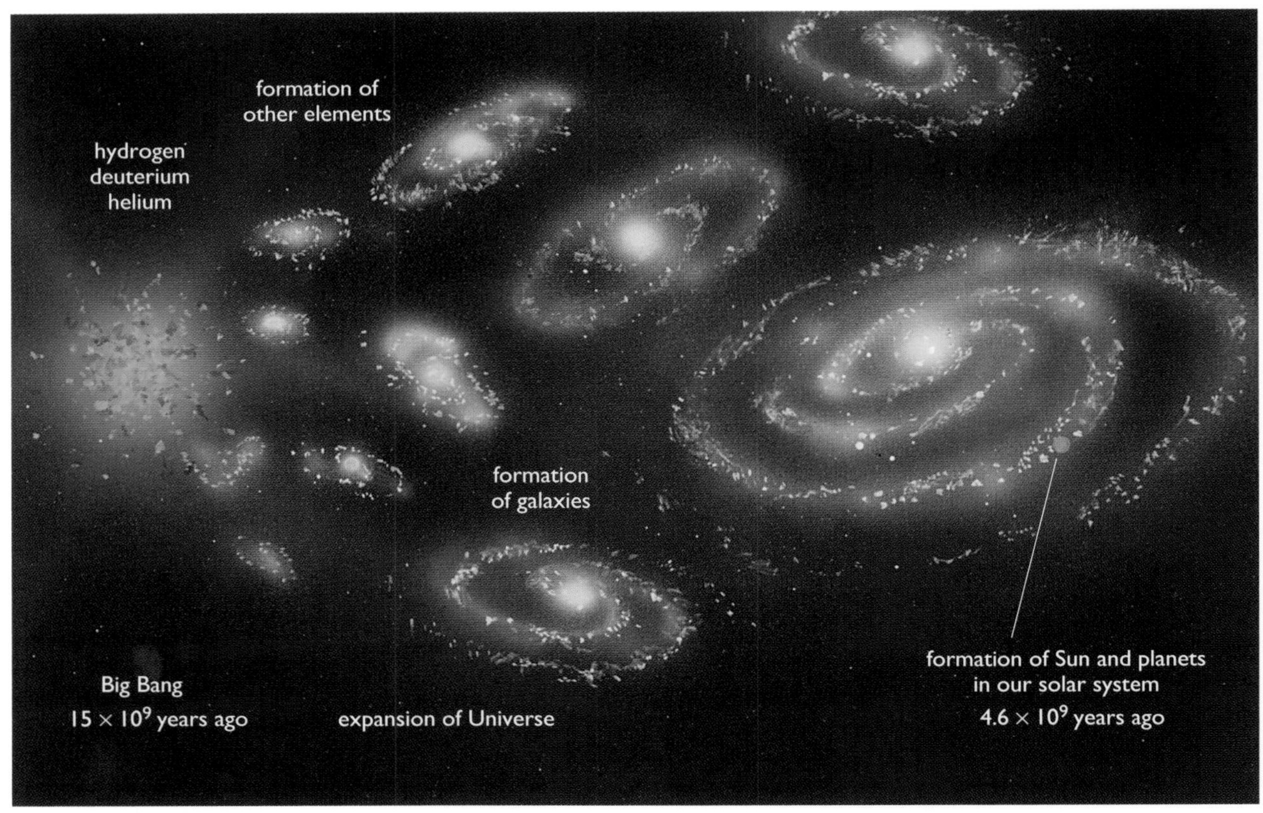

Figure 2.9
Schematic view of the evolution of the Universe, our galaxy and Sun from the Big Bang which occurred 15×10^9 years ago. As the Universe expanded and cooled, galaxies were formed. Our galaxy, shown on the right, is one of the billions of galaxies that exist in the Universe.

Box 2.4 Sizes and distances of astronomical objects

The Earth has an average diameter (it is not a perfect sphere) of 12 742 km (1.3×10^4 km), two orders of magnitude smaller than that of the Sun, which is about 1.4×10^6 km. The average Earth–Sun distance of 1.5×10^8 km is used as the **astronomical unit** (AU) of distance. The distance of the Sun from the next nearest star is 4×10^{13} km, or nearly 3×10^5 AU. Furthermore, the Galaxy in which our solar system is located is 9.5×10^{17} km (about 6×10^9 AU) in diameter and contains 10^{11} (one hundred billion) stars. When you look up at the Milky Way, the band of bright objects that crosses the night sky, you are looking at the stars in the main part of our galaxy. The Universe as a whole contains billions of galaxies. The AU is minute when compared with another unit of distance, the *light-year*, which is the distance travelled by light in one year, and is used for interstellar and intergalactic distances: 1 light-year = 9.46×10^{12} km (about 6×10^4 AU).

Figure 2.10 shows one type of meteorite, known as a **carbonaceous chondrite**. This is a very important class of meteorite because, as the name implies, they contain 2–5% by mass of carbon in the form of organic compounds (of the kind introduced in Chapter 1); they also contain up to 20% by mass of water in chemical combination with these compounds and with silicate minerals.

The carbonaceous nature of these meteorites and the presence of chemically combined water in them is evidence that they have not experienced significant

Figure 2.10
A carbonaceous chondrite from the meteorite shower (a group of meteorite fragments, probably from the breakup of a larger body) that fell in Orgueil, France, in 1864.

heating (probably not above about 200 °C) since they formed, unlike the crystalline and metal-rich types (Box 2.3), and they are considered most likely to represent the average elemental composition of the Solar System as a whole. Since meteorites can be analysed in the laboratory, we can measure the abundance of the elements in them very accurately.

- Looking at Figure 2.7 and bearing in mind that the Solar System is composed of 'secondhand elements' (Section 2.3.1), where might we look for the principal 'archive' of Solar System composition?

- The Sun is by far the largest body in the Solar System, and we can surmise that it might be reasonably representative of the whole system, in terms of elemental composition.

We obviously cannot determine the chemical composition of the Sun directly. We could analyse the chemical composition of the particles liberated as the solar wind, but these do not actually reach the Earth's surface (Section 2.3.1). More reliable data come from spectroscopic measurements (spectral analyses) of the Sun's radiation; and while we are doing this, we can make spectral measurements of other stars too.

Spectroscopic measurements of stars within the solar neighbourhood (arbitrarily taken to be a region of space within about 3×10^{16} km, i.e. 2×10^8 AU, of the Sun) show that the chemical elements are always present in the same relative proportions as in the Sun. When we compare the elemental composition of the Sun with that of carbonaceous chondrites, we see a striking similarity (Figure 2.11). The agreement in results from meteorites, the Sun and other stars of our galaxy gives us some confidence that Figure 2.11 is a good representation of the average relative cosmic abundance of the elements.

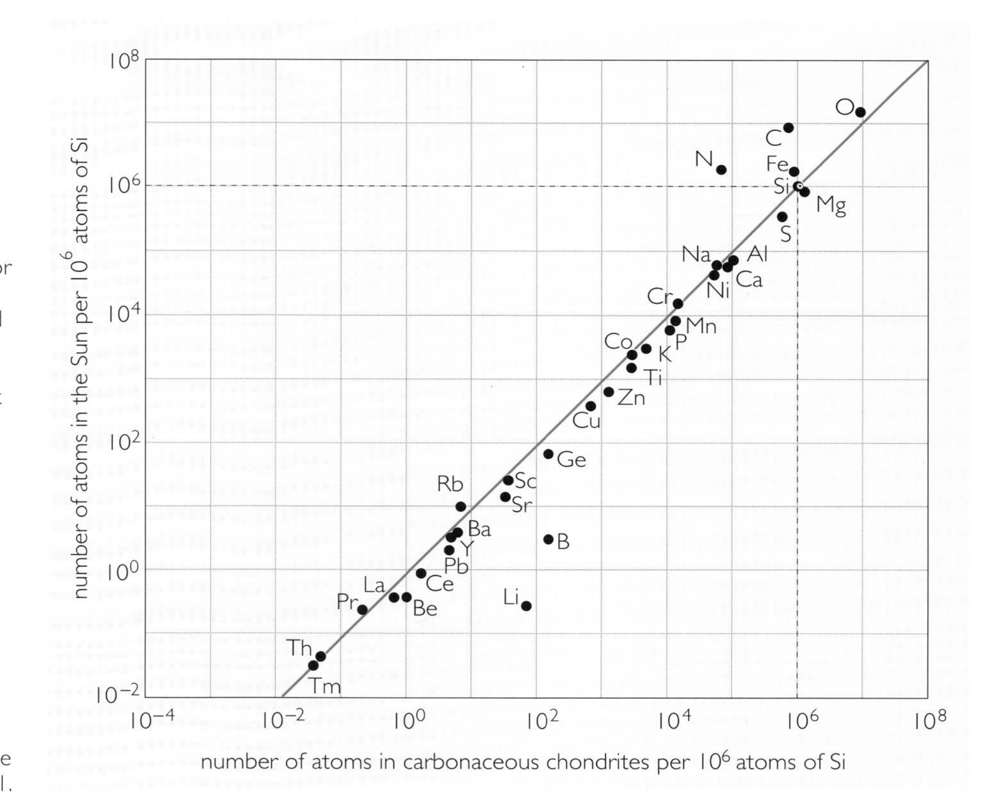

Figure 2.11
Relative abundance of elements in the Sun and in carbonaceous chondrites. For direct comparison, each element has been expressed as the number of atoms relative to a standard of 10^6 atoms of silicon. An element with exactly the same abundance in the Sun and carbonaceous chondrites would plot on the line shown. Note that the axes are logarithmic because of the enormous range of the numbers involved – each division represents an order of magnitude (factor of 10) difference in concentration, and 10^0 is equal to 1. If you are not familiar with the element symbols, refer to the Periodic Table in Appendix 1.

As an example of how to use the graph in Figure 2.11, you can see that for every 10^6 atoms of Si (silicon) in the Sun there are nearly 2×10^6 atoms of nitrogen (N), while for every 10^6 atoms of Si in carbonaceous chondrites there are about 8×10^4 atoms of N. This value for nitrogen lies further from the straight line than nearly all of the other elements in Figure 2.11. Overall, however, the correspondence between the Sun and carbonaceous chondrites is striking.

As you have seen, the processes responsible for producing the cosmic abundance of the elements all took place *before* the Solar System was formed. The Sun and the planetary bodies of the Solar System were created out of materials that were themselves manufactured in other, long-vanished stars. Some elements have been involved repeatedly in stellar formation, going in and out of stars as they were created, evolved and died. The Solar System represents a 'snap-shot' in time, recording the stage of chemical evolution reached in our galaxy 4.6×10^9 years ago; changes since that time have been relatively minor. Hydrogen is being used up as stars continue to generate energy and produce new elements, but the newly created elements are being produced in proportions that are generally similar to those in Figure 2.11, because the element-forming processes are those we described earlier (Figures 2.4 to 2.6).

Question 2.5
Hydrogen and helium are missing from Figure 2.11. Would you expect them to plot on the straight line?

About 99% of the mass of the Solar System is concentrated in the Sun, which has a mass of 2×10^{30} kg. As the Solar System can be thought of as a representative sample of the Universe, we can deduce that a truly 'average' cosmic composition is one dominated by hydrogen and helium.

- How can you tell from common knowledge that the Earth does not have a truly cosmic composition?
- The Earth's atmosphere is dominated by nitrogen and oxygen, and its chief repository of hydrogen is the oceans. Helium is not a gas you encounter on a daily basis.

The Earth is relatively depleted in elements like hydrogen and helium because of processes that took place during its formation. Most of the hydrogen and helium, along with other volatile elements and compounds (i.e. those that readily vaporize), either never accumulated on the planet in their cosmic proportions, or were lost as a result of heating events. Moreover, as you will see in Chapter 3, there is a different distribution of elements at the surface and in the interior of the Earth, because geological activity has acted to redistribute the elements. For instance, large-scale melting processes deep within the planet have formed a metallic core made predominantly from iron and nickel; and, as we have seen, the atmosphere is dominated by the elements nitrogen and oxygen, while the oceans are made of water (hydrogen and oxygen). What we *can* say, however, is that the starting materials from which the Earth and other planets formed were present in their cosmic proportions at the time of formation. The mass of the Earth (6×10^{24} kg), is only about one-millionth of that of the Sun, but Jupiter is about 300 times more massive than the Earth. Jupiter's strong gravitational field has caused it to retain its cosmic complement of hydrogen and helium, along with all of the other elements, including those necessary for life – indeed, in 1996 the Galileo space probe recorded that Jupiter's elemental composition is similar to that of the Sun. However, there is no life on Jupiter, in part because of the low temperature (it is a long way from the Sun, Figure 2.7) and its strong gravitational field. But the main reason is believed to be that the massive excess of hydrogen has 'mopped up' the carbon by forming methane, CH_4 (Figure 1.5a), leaving none available to form the complex organic molecules necessary for life.

Question 2.6

(a) The carbon in the Sun is said to be 'secondhand'. What is meant by this?

(b) We have on Earth samples of *solar* carbon available for study – how did they get here?

2.4 From interstellar dust to planets

If you look up at a clear night sky you may be struck by the fact that – away from the Milky Way – most of the sky is dark. This space between the stars is known as the **interstellar medium**, and you might conclude that it is empty. Indeed, up to a point you would be right, because interstellar space is an almost perfect vacuum. And yet, within our galaxy at least, this apparent emptiness contains about 10% of the total galactic mass (amounting to 2.2×10^{40} kg of interstellar matter). Clearly, space is not empty at all. So what *does* lie between the stars? Elements are ejected into interstellar space by the various processes described earlier, and the interstellar medium is thus a mixture of atoms, high-energy particles, molecules and accumulations of molecules in the form of interstellar or cosmic dust.

Let us now consider some of these materials in the interstellar medium, starting with the dust. Stars lose mass in stellar winds (Section 2.2.2), or explosively in nova outbursts and supernovae explosions, all of which disperse material widely through space. As hot stellar plasma is expelled and moves away from stars, the temperature and pressure within it drop, and eventually it condenses to form gases and, ultimately, solids. There are no liquids in space, because of the very low pressures, and so gases condense directly to solids – and when re-heated they vaporize directly to gases (see Box 2.5).

2.4.1 Interstellar dust

The dust particles that form by condensation around and between stars have a typical size range of 10^{-6} to 10^{-9} m. They are mostly composed of a variety of materials ranging from carbonaceous grains to oxides, such as magnesium oxide (MgO) or aluminium oxide (Al_2O_3) (and perhaps even simple silicates). We should stress here that the constituent *elements* (e.g. aluminium, Al, and oxygen, O) were synthesized by nuclear *fusion* in stars, but that the *oxides* and other compounds would be formed by *chemical reactions* between the elements, probably in the envelopes of stars where the gases are cooler. The carbonaceous grains may be *inorganic* or *organic* (see Box 1.2) – the carbon of inorganic carbonaceous grains is in elemental form (diamond or graphite), possibly also combined in silicon carbide (SiC). Organic carbonaceous grains may contain complex molecules such as polycyclic aromatic hydrocarbons.

Dust grains formed around stars can interact with other constituents of interstellar space. One important astronomical process involving dust grains seems to be their role as catalysts in the conversion of hydrogen (H) atoms into molecules (H_2). These

Box 2.5 Three states of matter

We use water as an example of the relationship between the solid, liquid and gaseous states of a substance. You don't need to understand all the details of what follows, just the main conclusion.

Figure 2.12 is a graphical representation of the different physical states of water with respect to *vapour pressure* and temperature. Note that vapour pressure refers to the pressure of water vapour within the air and not the total pressure of the air itself. For example, at 20 °C, the vapour pressure of water is only about 2300 Pa, compared with an atmospheric pressure of about 10^5 Pa (Box 2.2). In Figure 2.12, the three states of water are shown: solid (ice), liquid water, and water vapour. The curved lines are boundaries between these three states and they are labelled with the processes that cause changes of state across those boundaries.

Taking a familiar example, let us consider water in a frozen state, for instance some ice cubes from the deep-freeze, being warmed up to approximately 0 °C. At this temperature, the vapour pressure is about 611 Pa. This is called the *triple point*, and is a unique set of conditions under which the three phases of water can coexist in equilibrium. Under any other conditions, water only occurs in one or two phases in equilibrium.

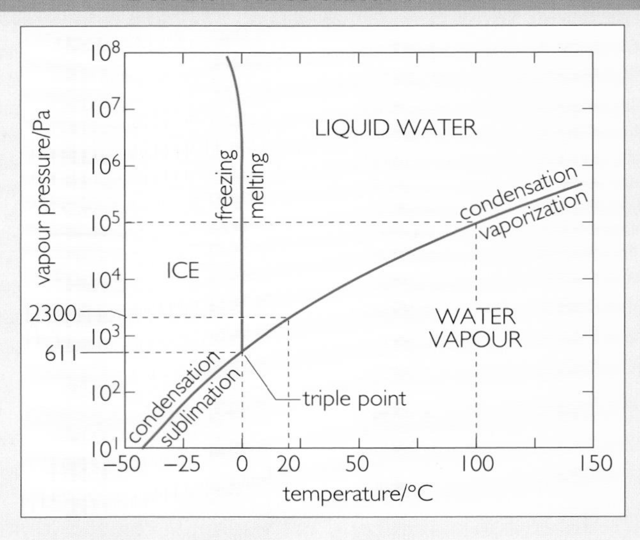

Figure 2.12
Graph of vapour pressure against temperature for the three states of water, showing boundary conditions between each state and the processes occurring across those boundaries. The triple point is very close to 0 °C (0.0075 °C) and 611 Pa. It represents the *only* conditions under which ice, liquid water and water vapour can coexist at equilibrium.

When the water is heated further, to 100 °C, its behaviour follows the line labelled vaporization/condensation on Figure 2.12 until eventually a vapour pressure of 10^5 Pa is reached. This is equivalent to atmospheric pressure, and so the water begins to boil.

Under the normal conditions of our everyday experience, water exists as liquid plus vapour, i.e. its behaviour follows the vaporization/condensation line in Figure 2.12 which is at the interface of these two states. However, if we change the conditions then we can get a variety of different outcomes. For instance, if we artificially keep the vapour pressure below 611 Pa then, upon heating from low temperatures, water will *sublime* directly from a solid to the vapour phase. An important consequence of moving across the boundaries shown in Figure 2.12 is that, depending upon the direction in which the crossing is made, *extra* energy over and above that needed simply to heat things up is either absorbed or emitted. This additional energy is known as *latent heat*. To change water from ice to a liquid and vice versa, *latent heat of fusion* is involved. Similarly, *latent heat of vaporization* is involved in the change from liquid to vapour – incidentally, latent heat is what causes severe burns if you accidentally expose skin to steam.

The reason we cannot actually see water evaporating is that at normal temperatures the process is relatively slow. However, molecule by molecule the liquid is eventually transferred to the vapour phase and escapes into the surrounding air. This is essentially what happens at the surface of the sea (or of a lake), particularly at low latitudes where temperatures are high. As the vapour rises, it eventually condenses out of the air and forms clouds. When the temperature falls below a certain point, the moisture falls to the ground as precipitation (rain, snow, hail).

- From Figure 2.12, which two states of water must coexist at the polar ice-caps at, say, −30 °C?

- At such low temperatures, only ice and water vapour can coexist, and ice sublimes to vapour, and condenses again to ice – there is no liquid water present.

Exactly the same kinds of relationships apply to other substances – though of course their temperature–vapour pressure diagrams would be different from Figure 2.12. Therefore the central conclusion is: at the very low temperatures and pressures that prevail in interstellar space, *all* gases must condense directly to solids, and sublime back into gases when heated. Hence there are no liquids in interstellar space.

hydrogen atoms may be veterans of the cosmic cycle, having been in and out of stars many times. But they may also be primordial products of the original Big Bang; in other words, processes in the interstellar medium can involve materials never previously involved in the formation of stars.

Hydrogen molecules can also be formed in interstellar space simply by interactions between hydrogen atoms and hydrogen ions, i.e. solely by reactions in the gaseous phase, without the intervention of dust particles. Of course, hydrogen is not the only element involved in such reactions. In fact, reactions between gases, and between gases and solids within the interstellar medium, result in ever more complicated molecules being produced, which then remain floating in space until they eventually stick onto the surfaces of **refractory** dust grains. These are oxides or other inorganic compounds that are resistant to heat, decomposition, melting and chemical attack, and they become progressively surrounded by both simple and complex molecules. Thus, as we shall see in the next section, the interstellar medium is like a giant chemistry laboratory, in which all kinds of compounds are synthesized and aggregated together, albeit rather slowly because of the very low temperatures and near-vacuum conditions.

2.4.2 Interstellar chemicals

About 1% by mass of the interstellar medium consists of dust. The rest is gas – mostly neutral atoms and molecules, but some ions as well – made up both of primordial components (left over from the Big Bang) and of materials expelled from stars. (~~This dispersed material is the 'dark matter' of the Universe that is so often discussed by astronomers and astrophysicists.~~) Most of the interstellar complement of carbon, hydrogen, nitrogen and oxygen remains in the gas phase, and only some resides in or on dust particles. These four elements are not only (along with helium) the most abundant in the cosmos (cf. Figure 2.11), they are also the elements most crucial to life – and the processes of interstellar chemistry have a profound effect on their distribution.

Matter is not uniformly distributed throughout the interstellar medium. Within our own galaxy, it is concentrated into regions known as *interstellar clouds*. There are two types of cloud: diffuse and molecular. The diffuse clouds contain only simple molecules, such as H_2, and have a density of about 10^6–10^8 particles per cubic metre of space. We will not consider them further here. The molecular clouds contain a variety of more complex molecules and have a much higher density (10^9–10^{12} particles per cubic metre). To appreciate the low densities of molecular interstellar clouds, consider for comparison the fact that with the help of the Avogadro constant (Box 2.1) we can estimate that a cubic metre of the Earth's atmosphere contains in the order of 10^{25} particles, nearly all in the form of molecules of nitrogen and oxygen.

Molecular clouds are typically 10^{13}–10^{15} km across. Table 2.1 shows the range and complexity of some of the compounds identified among the interstellar molecules in these clouds. Remember that, unlike elements synthesized in stars by nuclear fusion reactions, the molecules shown in Table 2.1 are the result of *chemical reactions* that took place within the interstellar clouds. You are not expected to understand (still less remember) the *details* of Table 2.1, but you should appreciate that both simple and complex molecules are represented.

Table 2.1 Simple molecules detected in molecular clouds in interstellar space. We have not shown all the compounds that have been identified and, in any case, new ones are being found all the time. The molecules given in **bold** type are those which are also detected in comets (see Table 2.2). You are not expected to memorize the details of this table.

simple hydrides, oxides, sulfides, halides and related molecules				
H_2	**CO**	**NH_3**	CS	$NaCl$
HCl	SiO	SiH_4	SiS	$AlCl$
H_2O	**SO_2**	CC	**H_2S**	KCl
	OCS	**CH_4**	PN	AlF
	HNO?			
nitriles, acetylene derivatives and related molecules				
HCN	**$HC\equiv C-CN$**	$H_3C-C\equiv C-CN$	H_3C-CH_2-CN	$H_2C=CH_2$
H_3CCN	$H(C\equiv C)_2-CN$	$H_3C-C\equiv CH$	$H_2C=CH-CN$	$HC\equiv CH$
$CCCO$	$H(C\equiv C)_3-CN$	$H_3C-(C\equiv C)_2-H$	HNC	
$CCCS$	$H(C\equiv C)_4-CN$	$H_3C-(C\equiv C)_2-CN$?	$HN=C=O$	
$HC\equiv CCHO$	$H(C\equiv C)_5-CN$		$HN=C=S$	
H_3CNC				
aldehydes, alcohols, ethers, ketones, amides and related molecules				
$H_2C=O$	**H_3COH**	**$HO-CH=O$**	H_2CNH	
$H_2C=S$	**H_3C-CH_2-OH**	$H_3C-O-CH=O$	H_3CNH_2	
$H_3C-CH=O$	H_3CSH	$H_3C-O-CH_3$	H_2NCN	
$NH_2-CH=O$	$(CH_3)_2CO$?	$H_2C=C=O$		
cyclic molecules				
C_3H_2	SiC_2	C_3H		
ions				
CH^+	HCO^+	$HCNH^+$	H_3O^+?	
HN_2^+	$HOCO^+$	SO^+	HOC^+?	
	HCS^+		H_2D^+?	
radicals				
OH	C_3H	CN	HCO	C_2S
CH	C_4H	C_3N	NO	NS
C_2H	C_5H	H_2CCN	SO	
	C_6H	$HSiCC$ or $HSCC$		

? is used to indicate uncertainty in interpretation of the data.
$C=C$ and $C\equiv C$ represent double and triple bonds, respectively.

Some of the simpler molecules shown at the top of Table 2.1 have names with which you may be familiar. We have already encountered H_2 and H_2O (hydrogen gas and water), CO is carbon monoxide, NH_3 is ammonia, and HCl is hydrogen chloride (which becomes hydrochloric acid when dissolved in water). These molecules are all either in the gas phase in the interstellar medium, or are present as solids (i.e. condensed gases) accreted onto dust grains.

The details of molecule formation in interstellar clouds are complex and we need note only that so-called *ion–molecule reactions* are the mainstay of interstellar chemistry in the near-vacuum of space. The predominant chemical species in interstellar molecular clouds are H, H_2 and He, and these are constantly being ionized through interaction with high-energy particles generated by explosive events such as supernovae. The ions so formed may be different from those encountered in 'normal' chemistry on Earth, and they can interact in many different ways with neutral molecules. For example, a hydrogen molecule (H_2) may be struck by a high-energy particle, which results in an electron being lost and the formation of an ion from the hydrogen molecule (H_2^+). This ion may then interact with another H_2 molecule to form an H_3^+ ion and a neutral hydrogen atom:

$$H_2^+ + H_2 \longrightarrow H_3^+ + H \qquad \text{(Equation 2.1)}$$

The resulting hydrogen atom may then interact with another to form a hydrogen molecule and so repeat the cycle. The H_3^+ ion can itself be involved in several kinds of reaction. Here are a few examples (which you are *not* expected to memorize):

$$H_3^+ + CO \longrightarrow HCO^+ + H_2 \qquad \text{(Equation 2.2)}$$

$$H_3^+ + N_2 \longrightarrow N_2H^+ + H_2 \qquad \text{(Equation 2.3)}$$

$$H_3^+ + O \longrightarrow OH^+ + H_2 \qquad \text{(Equation 2.4)}$$

$$H_3^+ + C_2 \longrightarrow C_2H^+ + H_2 \qquad \text{(Equation 2.5)}$$

The net effect of these reactions is to produce chemical species that are more complicated than the starting materials. Thus, in the case of Reaction 2.2, carbon monoxide (CO) is converted into HCO^+, which will eventually become methanal, H_2CO, (previously known as formaldehyde) after a further reaction involving hydrogen gas. The product from Reaction 2.3 will eventually lead to ammonia (NH_3), and that from Reaction 2.4 to water. Reaction 2.5 is the starting point for the formation of hydrocarbons, i.e. organic molecules. Notice how in Reactions 2.2 to 2.5 the H_3^+ has been converted back into H_2, essentially the starting material. The H_2 is then, of course, able to take part in further ion–molecule reactions (Equation 2.1).

The products of ion–molecule reactions in interstellar clouds remain in the gas phase until they accumulate onto grain surfaces. If part of a molecular cloud were to collapse and form a star, the high temperatures involved would convert any molecules formed previously by ion–molecule reactions back into their original elements. For instance, a methanal molecule would decompose back into H, H, C and O. This is another glimpse of the cosmic cycle in action.

When you consider that temperatures cannot be much above absolute zero (Box 2.2) and pressures approach those of a vacuum, the variety of organic compounds that can form in interstellar clouds (Table 2.1) is truly amazing. However, the very 'thin' nature of the clouds, coupled with the very low temperatures and pressures, precludes the formation of more complex molecules, especially those that are found in living organisms (Box 1.5). Before that can happen, the molecules must become less dispersed and higher pressures and temperatures must prevail. Only on the surface of a larger body like a planet might such an environment be found. However, as you will see in succeeding sections, a planetary surface may be a *necessary* condition for life to develop and evolve, but it is by no means automatically a *sufficient* condition.

2.4.3 An idea of the Marquis de Laplace

Astronomers have shown that stars are born in giant interstellar molecular clouds (Figure 2.13). The distribution of molecules within these clouds is not uniform. The clouds have more dense regions towards which gravity can draw matter from other less dense regions. Processes in these star 'nurseries' are much too slow for us to observe directly, and such regions are in any case difficult to observe because they are more opaque than elsewhere. Hence, building a model for stellar evolution depends on indirect evidence and a few lucky observations. One such piece of luck is the event captured in the photograph in Figure 2.14: the young star β Pictoris is surrounded by a disc of diffuse matter. Such an observation lends support to the idea, first formulated by the Marquis de Laplace in 1796, that the Solar System evolved from a rotating disc-like cloud of dust and gas, called the **solar nebula**, which condensed and accreted to form a number of centres of mass which became the Sun and planets. Figure 2.15 shows one view of this model, which has observational support, is flexible and forms a reasonable basis for discussion and elaboration.

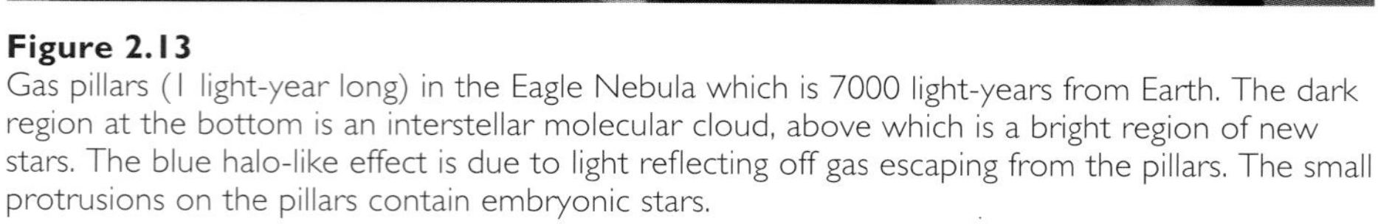

Figure 2.13
Gas pillars (1 light-year long) in the Eagle Nebula which is 7000 light-years from Earth. The dark region at the bottom is an interstellar molecular cloud, above which is a bright region of new stars. The blue halo-like effect is due to light reflecting off gas escaping from the pillars. The small protrusions on the pillars contain embryonic stars.

Figure 2.14
The young star β Pictoris is surrounded by a flattened disc of diffuse matter, illuminated by the star. This disc is thought to be the precursor to a planetary system.

Figure 2.15
Schematic illustration of a version of Laplace's theory for the origin of the Solar System. A cross-section of the rotating solar nebula is shown in the top picture to illustrate the thickness of the disc. Accretion of the nebula eventually resulted in the Solar System. (The orbits of the outer two planets, Neptune and Pluto, cross one another (Figure 2.7), but that is not shown in this simplified diagram.)

In brief, Laplace proposed that irregularities in the density of the solar nebula caused it to break up into many accreting small bodies, or **planetesimals** (with diameters up to about 10 km). The constantly changing gravitational fields during accretion thickened the disc as the orbits of the planetesimals were perturbed. Eventually, most of the planetesimals collided and accreted to larger centres of mass that became the planets, resulting in the system becoming more disc-like again.

The action of gravity in drawing matter together converts gravitational **potential energy** (that which you have when repairing the roof) into **kinetic energy** (that which you increasingly gain when you fall off), and ultimately into heat and other forms of energy when the matter collides with something else (when you hit the ground). So, local collapse of part of an interstellar molecular cloud to form the solar nebula generates heat, i.e. the temperature rises. While the cloud is still diffuse, this heat can be lost by radiation, but as its density increases, the nebula becomes more opaque. This allows more and more heat to be retained within the nebula, while the pressure also increases further as the collapse continues. This is how the nuclear fusion reactions in the Sun are believed to have been 'fired up', as it began life as the star at the centre of our solar system.

2.4.4 Formation of the planets

From previous sections, it should be clear that conditions were not uniform within the solar nebula. There was a temperature gradient from the highest temperature at the centre, where the Sun was formed, to the lowest temperature at the edge of the nebula, where it was effectively as cold as the rest of space. This radial decrease in temperature is reflected in a 'condensation sequence' of decreasingly refractory compounds (see Box 2.6) and hence in the nature of the planets (Figure 2.7). Towards the Sun we have four relatively small and dense objects: Mercury, Venus, Earth and Mars. These formed at relatively higher temperatures, which is why they are overall quite refractory, being composed largely of silicates and alumino-silicates (hence they are often called 'rocky planets'). In contrast, the four large outer planets are chiefly large balls of gas (Jupiter, Saturn) or gas with 'icy' interiors (Uranus, Neptune), because conditions were (and remain) colder in the region where they formed. Condensed gases are believed to dominate some of the satellites of the giant planets and possibly also the most distant planet, Pluto (but see Section 2.4.5).

Box 2.6 Condensation from the solar nebula

From Figure 2.11, the six most abundant *metals* in the solar composition are, in decreasing order (relative to 10^6 atoms of silicon, Si): iron (Fe), magnesium (Mg), aluminium (Al), sodium (Na), calcium (Ca) and nickel (Ni). The four most abundant *non-metals* are oxygen (O), carbon (C), nitrogen (N) and sulfur (S). This gives us a rough idea of the sorts of compounds that were likely to form as the solar nebula condensed, i.e. as it cooled in the near-vacuum of space (which means that condensation was direct from gas to solid, Box 2.5). We can summarize the main sequence of condensation of the principal elements and compounds with falling temperature as follows – note that temperatures are in kelvin (K):

above 1400 K: refractory oxides and silicates of aluminium, calcium and magnesium, with some other compounds and elements;

at about 1400 K: metal alloys, dominated by iron and nickel;

at about 1200 K: less refractory silicates of sodium and aluminium, with some other elements and compounds;

between 1000 K and 400 K: sulfides, dominated by iron sulfide, plus oxides (mainly of iron);

below 400K: various hydrocarbons, ammonia (NH_3) and water, in the form of sticky solids and ice. It is important to stress that the dominant elements hydrogen and helium remain largely uncombined; hydrogen is so abundant that nothing is left to combine with it; while helium is inert. Temperatures never fell low enough for hydrogen and helium to condense as solids, and so they either accreted to the growing Sun or remained in the nebula.

As temperatures at the margin of the solar nebula would have been very low, the bodies of the outer Solar System could simply be aggregations of interstellar material which never went through the heating and condensation processes that occurred close to the Sun. If so, anything in that region must logically be the remains of material that existed before the Solar System formed, i.e. interstellar dust, organic molecules, ice and other condensed gases. Could such materials have had any influence on the development of life on Earth? We consider this question in the next section.

- Do carbonaceous chondrites (Section 2.3.2) represent Solar System material that condensed from higher- or lower-temperature parts of the solar nebula?

- Carbonaceous chondrites are mixtures of high- and low-temperature components (Box 2.6) because they consist of silicate minerals with significant proportions of organic molecules which would have decomposed into their constituent elements (C, H, N, O) upon heating. So all we can say is that they cannot have experienced significant heating since they formed.

The organic constituents of meteorites were probably mainly formed on meteorite parent bodies after they had condensed from high temperatures. Although there is some evidence for higher-temperature synthesis of organic compounds such as polycyclic aromatic molecules (Section 2.4.1), compounds such as amino acids were almost certainly synthesized much later and at quite low temperatures, *after* the solar nebula had condensed to form planetesimals (Figure 2.15).

2.4.5 The Oort cloud and the Kuiper belt

Comets are kilometre-sized bodies composed essentially of aggregations of ice, interstellar dust and organic molecules. They also include frozen carbon monoxide and carbon dioxide, and so it will perhaps come as no surprise that scientists refer to comets as 'dirty snowballs'. Comets have orbits very different from those of planets (Figure 2.7). When a comet's orbit takes it closer to the Sun, it becomes heated.

(a)

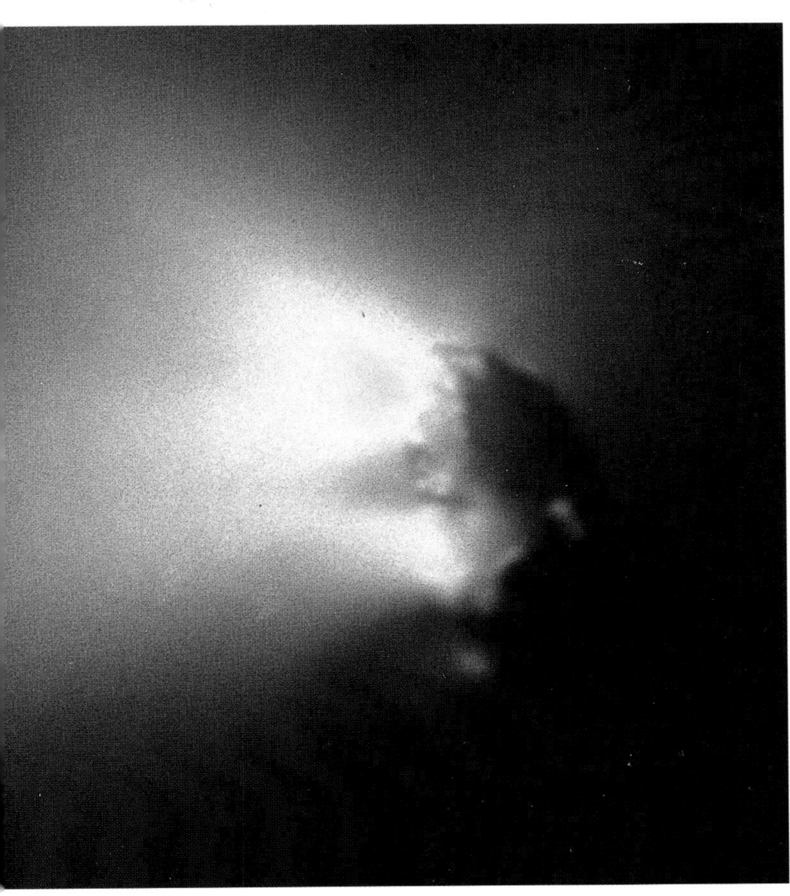

(b)

Figure 2.16
(a) Image of Halley's comet from the 1986 appearance. Its appearance in 1910 was more spectacular. Note that the 'tail' is not aligned along the comet's orbit, it always points away from the Sun. (b) A close-up view of the comet's head taken by the European Space Agency mission Giotto.

- What do you think is likely to happen to the ice and other condensed gases in the comet when this happens?
- Because of the low pressures in space, these condensed solids *sublime* directly to gases (Box 2.5).

Since comets are mixtures of condensed gases and dust, both components are lost during the heating. Sunlight scattered from the blown-off matter hides the solid body at the comet's head, and enables us to see the comet's characteristic 'tail' (Figure 2.16). If the Earth happens to cross the comet's orbit, some of the dust particles enter the atmosphere and can be seen as numerous shooting stars in the night sky – these are *not* the same as shooting stars from asteroidal dust (Box 2.3). The orbital paths of some comets are such that the Earth crosses them at the same time each year, and the time of the resulting *meteor shower* (or meteor stream) can be predicted. For example, the Perseid meteor shower (so-called because the meteors seem to come from the constellation of Perseus) occurs on August 12, and on a good night (with good viewing conditions) you can see 40 shooting stars per hour. They are the dust from the comet Swift–Tuttle, a short-period comet, with a period of about 130 years. In November the Leonid meteor shower (named for the constellation of Leo, from where it appears to emanate) can be seen. It is produced by dust from the Tempel–Tuttle comet, which has a 33-year period. Meteor showers are associated with myth and legend, for they were seen long before they were understood. The Taurid shower, for example, is thought by some historians to have sparked off the building of Stonehenge.

Where do comets come from? Some of them originate in the **Oort cloud**, a spherical shell of 'icy' bodies (estimated to total between 10^{11} and 10^{13}) that surrounds the Solar System at a distance of up to 10^5 AU, i.e. far beyond the orbit of Pluto. It is believed to be the source of the so-called *long-period* comets, which appear without having been documented before, or at least are not observed at regular intervals (that means in practice intervals longer than about 200 years). These comets are probably extracted from the Oort cloud by gravitational perturbations resulting from close encounters with nearby stars and molecular clouds, and they must have highly eccentric elliptical orbits – a possible long-period comet is illustrated in Figure 2.7. An example is Comet Hyakutake which approached within 15 million kilometres of Earth in early 1996 and attracted some publicity. Its period is estimated to be about 9000 years. Another is Comet Hale–Bopp which is due to be visible in 1997.

Of greater interest to us are the *short-period* comets, of which perhaps the most famous is Halley's comet, whose orbital period is 76 years. The Swift–Tuttle and Tempel–Tuttle comets, mentioned above, are just two examples of many others. Comet Shoemaker–Levy 9 was also a short-period comet, one that had broken into many fragments which subsequently crashed into Jupiter in July 1994, causing something of a public sensation (Figure 2.17).

Short-period comets probably come from the **Kuiper belt**, which consists of about 10^5 'icy' bodies that orbit the Sun at distances of between 30 AU (just beyond the orbit of Neptune) and 100 AU. It has also been suggested that Pluto is not a planet at all but just a large body in the Kuiper belt. The same may be true of its satellite (Charon) and of Triton, one of Neptune's satellites.

Comets have been repeatedly ejected from the Kuiper belt through gravitational perturbations by other astronomical bodies, just as long-period comets are ejected from the Oort cloud, and meteorites and asteroids are ejected from the asteroid belt. The astronomical bodies that perturb bodies in the Oort cloud and the Kuiper belt cannot be planets of the Solar System (Figure 2.7), they must be outside it. At which point it is salutary to remind ourselves that the Solar System is 'but a flea on the elephant' of the Galaxy (see Box 2.7), and also that our galaxy is but one of billions in the Universe (Box 2.4).

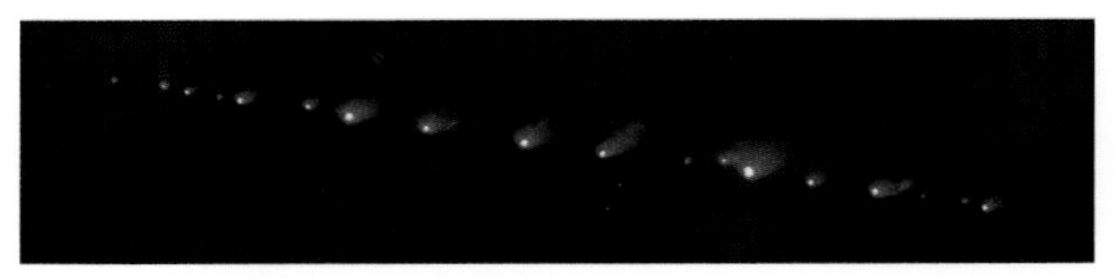

(a)

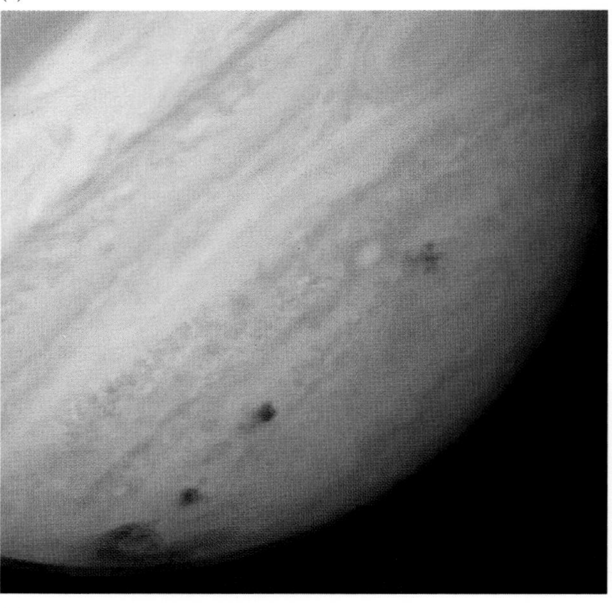

(b)

Figure 2.17
(a) Fragments of Comet Shoemaker–Levy 9. (b) Impacts of Comet Shoemaker–Levy 9 fragments on Jupiter, July 1994. (This photograph was taken from the Hubble Space Telescope.)

Box 2.7 Our place in the Galaxy

The Sun is one 'common-or-garden' star in an ordinary galaxy. We see our galaxy as an edge-on, inside view of a disc-shaped aggregation of stars that appears to encircle the Earth (the Milky Way, Box 2.4). Astronomical observations show that the Galaxy is of the spiral-arm type (Figure 2.18), and variations in brightness of the Milky Way, depending on the direction in which we look, indicate that the Sun is located on the outer part of one of the arms.

Our galaxy is not a fixed object. The stars within it orbit around its centre, the galactic centre, roughly in the plane of the disc, i.e. in the galactic plane. Their motion is not simple because, as a glance at Figure 2.18 will show, matter is not uniformly distributed in galaxies. So a body moving through the Galaxy is subject to continually changing gravitational fields, associated with variably dense groups of stars or interstellar clouds (Section 2.4.1). The mass distribution changes as the star groups and clouds move relative to one another, and the Solar System moves through a constantly changing scene of varying gravitational fields. The Solar System goes around the galactic centre about once every 250 Ma (250 million years).

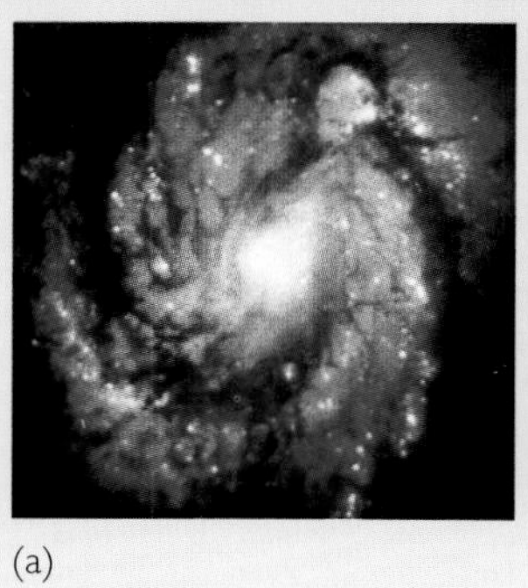

(a)

(b)

Figure 2.18
(a) 'Downward' view of a typical spiral-arm galaxy, Galaxy M100.
(b) Edge-on view of the Andromeda Galaxy.

Question 2.7

Comets contain water (see Table 2.2). In general terms, how and where was the water in comets likely to have been formed? How did this water get into comets?

Table 2.2 Simple molecules detected in comets. You are *not* expected to memorize the details of this table. The molecules given in **bold** type are those which are also detected in molecular clouds (Table 2.1).

simple hydrides, oxides, sulfides, halides and related molecules			
H_2O	**CO**	**NH_3**	CS_2
	SO_2	**CH_4**	**H_2S**
	CO_2		
	OCS		
nitriles, acetylene derivatives and related molecules			
HCN	**$HC{\equiv}C{-}CN$**		
aldehydes, alcohols, ethers, ketones, amides and related molecules			
$H_2C{=}O$	**H_3COH**	**$HO{-}CH{=}O$**	
$H_2C{=}S$	**$H_3C{-}CH_2{-}OH$**		
simple molecules			
He	S_2		
Ar	N_2		

$C{=}C$ and $C{\equiv}C$ represent double and triple bonds, respectively.

Short-period comets come sufficiently close to Earth to be investigated in detail. Most of the molecules so far found in comets (Table 2.2) have also been detected in molecular clouds (Table 2.1), and it seems likely that comets from the Kuiper belt (*and* those from the Oort cloud) consist of primitive materials – i.e. they are remnants of the dense molecular cloud from which the Solar System was formed – and they probably contain a cosmic complement of all elements except hydrogen and helium. Indeed, they may represent the kinds of bodies that aggregated to form the outer planets and satellites of the Solar System. We can envisage a possible process. Refractory dust grains built up coatings of condensed gases, trapping some organic molecules in the process. More complicated organic molecules were then produced as these composite grains were irradiated by ultraviolet radiation from nearby stars. Individual grains then began to stick together forming clumps, and the planetesimals began to take shape. When the Solar System eventually formed, materials not incorporated into the Sun or the planets remained unscathed and are mostly now resident in the Kuiper belt or the Oort cloud.

- Compare Tables 2.1 and 2.2. According to these tables, have hydrogen, helium, nitrogen and carbon dioxide been detected in both molecular clouds and comets? Is this to be expected?

- Hydrogen (H_2) is listed as occurring in molecular clouds but not in comets, and the reverse applies to helium (He). Nitrogen (N_2) and carbon dioxide (CO_2) are listed for comets but not for molecular clouds. This is somewhat unexpected, because comets are considered to be aggregates of interstellar dust and condensed gases, so these rather common substances should appear in both tables. That they do not may be due more to limitations of equipment and conditions of measurement than to actual absence of these molecules.

Cometary dust grains do not always vaporize or melt during atmospheric entry and some survive to be sampled. By means of specialized collection techniques aboard high-flying aircraft, completely unmelted cometary dust grains can be collected, with their original textures preserved (Figure 2.19). They have a characteristic 'fluffy' appearance – the pore spaces would have originally been filled with ice and condensed gases, which evaporated (sublimed, Box 2.5) during atmospheric entry, helping to carry heat away from the grains thereby allowing preservation of their delicate textures.

Figure 2.19
Scanning electron microscope picture of a cometary dust grain. The particle is about 10 μm across.

- Why should we have to rely on observations of cosmic dust particles for our investigations of comets? Could we not find a sample that reached the Earth's surface, just as many meteorites do?

- The principal reason is that comets are much less 'solid' than meteorites. Their condensed gases start to vaporize as soon as they get anywhere near the Sun, and the process continues, along with destruction of their organic components, when they reach the Earth's atmosphere. This leaves only the refractory dust grains, which can be collected by high-flying aircraft and from samples of Antarctic ice.

However, if a comet is large enough to survive total vaporization and fragmentation *en route* to the Earth's surface, it can pack a stronger punch than one would expect from anything so insubstantial as a 'dirty snowball'. On 30 June, 1908, a tremendous explosion rocked the inaccessible forests along the Tunguska River in Siberia. Trees were broken like matchsticks more than 50 km away from the blast centre. People were knocked over and window panes blown in 150 km away. All over northern Europe, for the whole of the next week, sunsets seemed peculiarly long and beautiful because of the pall of dust that had been shot into the upper atmosphere. What had caused the explosion long remained a mystery. At the site itself, scientists could find no sizeable crater and no fragments except for minute, fused pellets – apparently of terrestrial origin – embedded in the ground. Finally, in 1960, a thorough investigation was conducted by the Committee on Meteorites of the Soviet Academy of Sciences. They concluded that the explosion had been caused by the icy head of a comet, a few hundred metres in diameter and with a mass of about a million tonnes. This is much smaller than most comets seen roaming the Solar System (including Shoemaker–Levy 9). It now seems more likely that the absence of a crater is because the Tunguska event resulted from *explosion in the atmosphere* of a comet not more than 50 m in diameter. In fact, explosions of small comets in the upper atmosphere are now detected quite frequently thanks to the development of 'star wars' technology.

About 3×10^8 kg of extraterrestrial material are added to the Earth each year, comprising cometary and interplanetary dust, and meteorites (including carbonaceous chondrites, Figure 2.10). We know that these materials contain organic compounds. So, if extraterrestrial organic compounds are raining down on the Earth today, why not in the past as well? Could materials of this nature once have 'seeded' organic compounds on the sterile surface of the early Earth? The meteorites and cometary dust particles we can collect and study today may contain clues to the ancestral cosmic material from which life evolved. In Chapter 3, we examine (among other matters) the effects of meteorite impacts on the Earth, and in particular the nature of the early bombardment which helped to shape the planet and to form the Moon. Subsequently we look in more detail at how life on Earth might have started – comets and meteorites are implicated there too.

2.5 Summary of Chapter 2

1 The simplest element of all is hydrogen, the starting material from which all other elements were formed. The hydrogen itself originated when the Universe was born in the Big Bang. The lightest elements are formed by nuclear fusion of hydrogen and deuterium nuclei to produce helium (with the liberation of energy according to the equation $E = mc^2$). This process characterizes smaller stars like our own Sun, which still consists largely of unused hydrogen.

2 A feature of 'layered' stars is the CNO cycle, in which hydrogen nuclei interact with carbon, nitrogen and oxygen nuclei producing helium, and making other elements (outside the CNO cycle). Elements such as carbon and oxygen are made in larger and hotter stars by helium fusion. Fusion of these elements with more helium or with each other produces heavier elements (up to iron), often resulting in explosive nova and supernova events. Elements heavier than iron (up to uranium) are formed by neutron capture.

3 Some of the elements formed within stars are expelled into space as stellar winds, others remain in the body of the star until it eventually contracts and explodes. All eventually become available for recycling into new stars, where they participate in further fusion reactions with (mainly) hydrogen and helium, to form more new elements, which in their turn become available for participation in further nuclear reactions. Our own solar system (Sun and planets) is part of this cosmic cycle and was formed from 'secondhand' elements that were already in existence. Helium is the principal product of the Sun's own nuclear fusion. Virtually all of the Sun's other elements were formed in other stars prior to its birth.

4 The age of the Solar System (about 4.6 billion years) is known from measurements on meteorites. Chemical analyses of meteorites, along with spectral measurements of stellar radiation, provide information about the cosmic abundance of the elements.

5 The interstellar medium consists of dust and gas. This matter is not evenly distributed but tends to be concentrated into huge interstellar clouds, billions of kilometres across. Denser regions of such clouds attract matter from less dense regions (due to their gravitational field) and can evolve into solar nebulae, where dust, condensed gases and organic molecules can aggregate to form progressively larger bodies, reaching tens of kilometres across, called planetesimals.

6 Gravitational collapse of large interstellar clouds increases both the energy of the system and the pressure at its centre, where temperatures eventually rise enough for fusion reactions to start, whereupon a star is born. Our solar system is believed to have originated from a rotating disc-like molecular cloud (nebula). The planets grew by accretion of dust and planetesimals at various points down the temperature gradient away from the centre, where

temperatures were highest. The four inner (rocky) planets – along with the asteroids – formed nearest the Sun and are dominated by (refractory) silicates and metals. The outer planets, formed at lower temperatures, have much higher proportions of condensed gases and of organic compounds.

7 The materials of the planets were physically and/or chemically modified during gravitational collapse and accretion. Thus they no longer truly represent the primordial matter from which the Solar System was formed. Beyond the orbital limit of the planets, in the Kuiper belt and Oort cloud, are billions of orbiting bodies ranging in size from dust to planetesimals, and formed of aggregates of dust, condensed gases and organic molecules. These two regions are believed never to have been heated during Solar System formation and so probably consist of primordial matter. They are the source of comets which episodically enter the Solar System and occasionally collide with planets. Complex organic molecules in comets could have 'seeded' the Earth during and soon after its formation.

Now try the following questions to consolidate your understanding of this chapter.

Question 2.8
Consider a hypothetical star containing only hydrogen, some helium, and a little carbon. Why might we expect such a star to have some nitrogen and oxygen after the passage of time?

Question 2.9
Consider a star formed very early in the history of the Universe and composed only of elements formed during the Big Bang. (a) Could planets form round such a star, and (b) if so, would such planets be capable of supporting life?

Question 2.10
Life may have evolved on the planet Mars at some point during the history of the Solar System, but as far as we know (Chapter 1) there is presently no life on Mars.

(a) How old is Mars?

(b) How does Mars differ from Jupiter, other than in size?

(c) Would you expect the overall chemical composition of Mars to approximate the cosmic abundance of the elements?

Question 2.11
Dust particles and grains in the interstellar medium may be composed of elements such as aluminium and oxygen, in the form of the oxide Al_2O_3. What is the fundamental difference between the formation of the individual elements and the formation of these oxide grains?

Question 2.12

(a) How do asteroids differ from comets?

(b) Why does the Solar System have rocky inner planets and gaseous outer planets?

Question 2.13
The age of the Universe is given as 15×10^9 ($\pm\ 5 \times 10^9$) years. What therefore are the possible maximum and minimum figures for the age of the Universe?

Chapter 3
The early Earth – conditions for life?

In the late 1960s, Vic MacGregor, a geologist with the Greenland Geological Survey, worked out the relationships between different types of rock exposed at a place called Isua in western Greenland. The rocks included ancient sediments and volcanic lavas which had been subjected to complex folding by Earth movements and subsequently intruded by a mass of granite (Figure 3.1a). MacGregor knew that the sediments and lavas must have been older than the granite, possibly much older. They turned out to be the oldest rocks in the world, with an age of 3.8×10^9 years (i.e. 3.8 Ga).

(a)

(b)

Figure 3.1
(a) A specimen of 3.8 Ga old rock from Isua, western Greenland. (b) Outcrop of some of the oldest rocks, found in Canada. These are about 4 Ga old.

Since then, still older rocks have been found in Western Australia and northern Canada, dated at between 4.1 and 4×10^9 years (Figure 3.1b). Oldest of all are crystals of the mineral zircon, found in ancient sediments in Western Australia and dated at around 4.2×10^9 years. However, rocks older than those at Isua are less varied (though they do include ancient lavas and granites and appear also to have been subjected to complex folding). So the 3.8 Ga old rocks from Greenland provide us with the best information about what conditions may have been like on the early Earth.

The Isua rocks are the remains of sediments and lavas that did not differ fundamentally from the sediments and lavas of modern times. Among the original sediments were limestone and sandstone deposited under water, and the rocks contain traces of carbon which could be interpreted as the remains of life-forms (see Section 3.2.2). Limestones – calcium carbonate sediments – are nowadays typically formed by organisms, but they can be of inorganic origin also. The associated lavas in the Isua rocks have what is called a *pillow structure*, which indicates that they cooled under water.

3.1 The story in the oldest rocks

The first major inference that can be drawn from the oldest rocks is that the geological processes and cycles we recognize today were operating on Earth at least 4×10^9 (4 billion) years ago, albeit with some significant differences of detail.

Here is some of the evidence upon which that inference is based:

1. Sediments (including sandstones and limestones) and lavas were deposited under water, so there must have been bodies of liquid water at the Earth's surface, possibly ocean basins.
2. To form such sediments, land areas (albeit small) would have needed to be exposed to weathering and erosion, the products of those processes being transported to form the sediments in our (inferred) ocean basins.
3. The land areas being weathered and eroded could have been broadly similar to the upper parts of present-day continental areas, which approximate to granite in composition. Granite is a common rock that contains a good deal of quartz and yields quartz-rich sand when weathered and eroded.
4. The sediments were subjected to complex folding prior to intrusion by the granitic rock, and subsequently uplifted to form land that was then subject to further cycles of weathering and erosion.
5. Remains of possible life-forms in the sediments are also consistent with the presence of bodies of water at the Earth's surface, possibly of oceanic dimensions. The oldest rocks tell us little about what the atmosphere of the time was like. However, there is indirect evidence that it was probably rich in carbon dioxide and nitrogen, and devoid of oxygen.

These inferences are supported by the discovery in Western Australia in 1995 of the oldest preserved terrestrial landscape so far known. It is an area of continental crust that was being weathered and eroded about 3.5 billion years ago, before subsiding and being overlain by lavas with some shallow-water sediments (which included carbonate and sulfate minerals evidently precipitated from evaporating seawater). Such ancient (fossil) erosion surfaces preserved beneath overlying sediments are known as *unconformities*, and this discovery is proof that areas of continental crust were already above sea-level at a very early stage in the Earth's geological history.

Question 3.1

We can be fairly confident that there were no land-based life-forms on the early Earth. Bearing in mind what you read in Chapter 1, how is lack of atmospheric oxygen consistent with an absence of life on Earth?

The answer to Question 3.1 does not tell the whole story – in fact it seems to contradict what we said in point 5 above. As you will see later, many organisms on the present-day Earth live without oxygen – indeed, oxygen is toxic to them – and the earliest life-forms are believed to have been organisms of this kind. Lack of atmospheric oxygen may indeed be *consistent* with absence of life, but is not *proof* of its absence.

We may not be able to find direct evidence of the composition of the atmosphere on the early Earth, but the existence of very ancient sediments does enable us to be certain that temperatures at the Earth's surface must have been within a few tens of degrees of those prevailing today.

- How can we be so certain?
- For weathering, erosion and deposition of sediments to occur, there must have been water vapour in the atmosphere to form rain, which collected into rivers and ran off into the sea carrying weathering products with it. If there was liquid water at the Earth's surface, then average temperatures must have been greater than 0 °C and less than 100 °C.

This answer assumes that mean atmospheric pressure was similar to that of the present day. However, even if it were greater by a factor of (say) two, water would still freeze at only a few degrees below 0 °C, and would boil at only a few tens of degrees above 100 °C (Box 2.5). We know that the interior of the Earth was significantly hotter than it is today, and that average surface temperatures on the early Earth were probably somewhat higher than at present, as we shall see shortly. We may infer from this that rates of geological processes were somewhat greater than they are now, i.e. things happened faster then.

3.1.1 Structure of the early Earth

The second major inference that we can draw from the evidence in the oldest rocks builds upon our first (that geological cycles were in operation 4 billion years ago). It is that the Earth's internal structure was much as we believe it to be today, as summarized in Figure 3.2. This deduction is based on the probability that the Earth's outermost layer (the *crust*) was not greatly different from what it is now. The important point is that the Earth had acquired its layered structure by 4 billion years ago.

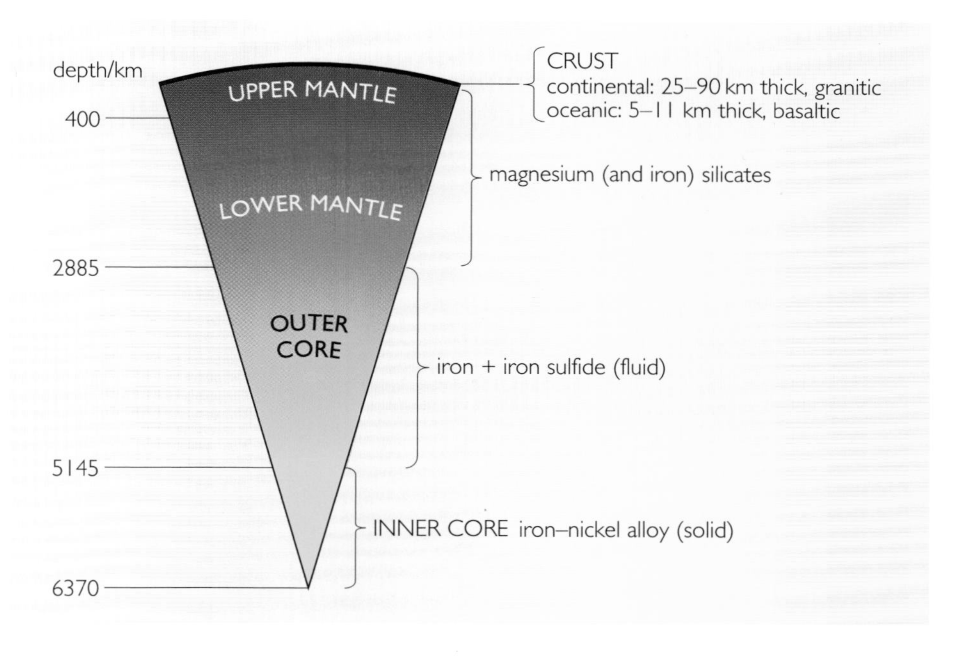

Figure 3.2 A schematic slice through the Earth, showing the major compositional features of the principal layers. *Note*: Strictly speaking, while the *upper* continental crust approximates to granite in composition, the continental crust *as a whole* has a composition intermediate between granite and basalt. For the purposes of this book, however, 'granitic' is a fair approximation.

The main points to note from Figure 3.2 are:

1. The core regions of the Earth are inferred to be of iron–nickel alloy, iron and iron sulfides (though other constituents have been proposed for the outer core), partly because of evidence from earthquakes (seismology) and partly on the basis of evidence provided by the various types of meteorite summarized in Box 2.3.

2 The composition of the mantle is also inferred from seismological data, as well as from meteorites, and from rock samples brought up from below the crust by volcanoes.

3 The (mainly granitic) continental crust is both thicker and *less dense* than the (mainly basaltic) oceanic crust, and it is chiefly for this reason that continental crust forms land areas, while oceanic crust underlies the ocean basins. It is probably not stretching credulity too far to suggest that if granitic continental crust was being eroded to deposit sediments in ocean basins 4 billion years ago, then those ocean basins were underlain by oceanic crust that was at least generally similar to the basaltic rocks that form the floors of our present-day oceans.

3.1.2 Early plate tectonics?

Given a crustal structure similar to that prevailing today, it seems reasonable to suppose that some sort of **plate tectonics**, the mechanism that drives *continental drift*, was operating 4 billion years ago. Plate tectonics continually recycles oceanic crust back into the mantle at subduction zones (Figure 3.3), and continually regenerates it at ocean ridges by the solidification of newly generated basaltic **magma**, some of which is erupted as lava on the sea-bed, to form the upper layers of the oceanic crust.

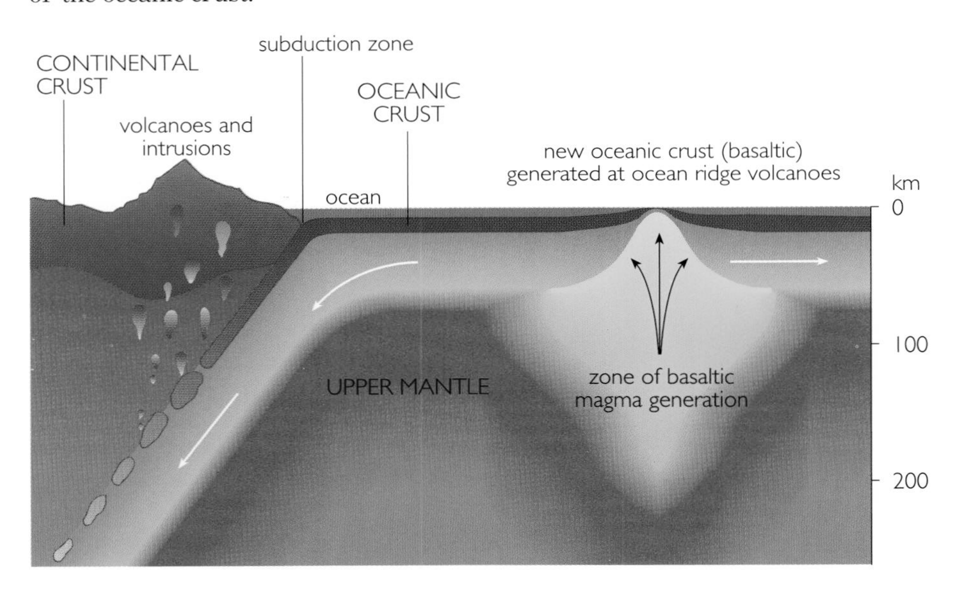

Figure 3.3 Schematic cross-section of the present-day plate tectonic cycle, showing old oceanic crust descending (being *subducted*) back into the mantle where it meets continental crust, and new oceanic crust being generated from basaltic magma at ocean ridges.

We know that modern internal Earth processes are mainly driven by heat energy released by the radioactive decay of unstable isotopes (Box 2.1), notably those of potassium (^{40}K), uranium (^{238}U) and thorium (^{234}Th). For any planet to remain stable, this internal heat has somehow to be lost to space by thermal radiation from the surface. That can only happen if the heat is transferred upwards to the surface, and the heat transfer takes place in two ways.

One process is at the atomic or molecular level: the heat energy is transferred from hot parts to those which are cooler, establishing a temperature gradient from hot to cold, and the heat 'flows' down the gradient. This is the phenomenon of **conduction**, which we experience frequently, for example, when a spoon is left in a cup of tea and its handle gets hot. Because silicate rocks are poor conductors, however, the Earth's internally produced heat cannot get to the surface fast enough

Figure 3.4
Highly schematic representation of simple convection cells, represented by arrowed flow lines. Hotter material is transferred from below upwards and cooler material from above downwards.

by conduction. Something else happens: materials are physically moved round by **convection**, the process we can see, for example, in a pan of boiling soup – except that inside the Earth it happens much more slowly. Under high pressures and temperatures, most solids can behave as highly viscous fluids, given enough time. When rocks are heated, they expand and their density decreases, making them buoyant relative to cooler and denser rocks. So, while hot rocks slowly rise in some regions, cooler rocks slowly descend in others, in a system of convecting cells rather like those shown in Figure 3.4, but considerably less uniform and regular. Convection is a much more efficient way of shifting excess heat than conduction, and virtually every part of the inner Earth is thought to be involved in convection, including the core.

- Can you suggest a reason why rates of convection in the early Earth would have been greater than they are today?

- Because of radioactive decay, the total amount of unstable isotopes decreases with time. So there would have been more radioactive isotopes decaying to heat the early Earth's interior than there are now, so there would have been more heat to lose and rates of convection would have been greater.

In fact, about five times more heat was being produced in the Earth's interior 4 billion years ago than there is today, but this was only partly because there were more radioactive heat-producing elements within the Earth at that time. Another factor was the early heating that the Earth received as the kinetic energy of the planetesimals was converted into heat during accretion (Section 2.4.3). In addition, an enormous amount of gravitational potential energy was released as heat when huge amounts of iron and nickel sank to the middle of the Earth, forming the core, probably about 4.5 billion years ago (see Section 3.4). Both of these components of the Earth's early or *primordial heat* continue to supply a significant fraction of the Earth's internal heat energy. All in all, then, convection would have been a good deal more vigorous 4 billion years ago, and the convection cells correspondingly smaller; so the rate of recycling of oceanic crust would accordingly have been greater.

Intuitively you might suppose from Figures 3.3 and 3.4 that regions where new magma is being formed (at *ocean ridges*) would be above the rising limbs of convection cells, and regions where older crust is sinking back into the mantle (called *subduction zones*) above the descending limbs. Alas for intuition, this appealing correlation has, for some years, been known to be invalid. The relationship between convection cells and plate tectonics is not so simple, though we need not go into details here.

However, we must mention two other important aspects of plate tectonics. The first is that as oceanic crust descends back into the mantle at subduction zones, some of it melts and the products rise as magma towards the surface, to contribute to the growth of continental crust (Figure 3.3). The second point is that because continental crust is less dense than oceanic crust, it has greater buoyancy and so it *cannot be recycled back into the mantle*. It therefore accumulates on the surface, and in fact, the area of continental crust has been growing for the last 4 billion years, though the rate of growth is believed to have decreased exponentially with time. (The area of continental crust had probably nearly reached its present extent by the time life finally began to emerge *onto land*, some 400 million years ago.)

The region where oceanic crust meets continental crust and starts to descend (the top of the subduction zone, at the left of Figure 3.3) is also a region where sediments that have accumulated on oceanic crust can become strongly folded (because they, too, are mostly not dense enough to be subducted). It is possible that the sediments in our oldest rocks were folded in such zones. This process has a bearing on the growth of continental crust: material eroded from the continents is eventually returned to them by accretion at the edges of the continental crustal layer, while new material is added from below (see left-hand side of Figure 3.3).

3.1.3 An interim summary

But what has all this got to do with Earth and life? So far in this chapter, we have been setting the scene for life to develop. Let us summarize: the possible traces of organisms in the 3.8 Ga old Isua sediments suggest that conditions at the Earth's surface might by then have reached a state where life could be sustained. If the inferences we have drawn from the oldest rocks are any guide, then by 4 Ga ago the Earth was already rather like the Earth we know today. Thus, it had a core and a mantle, an oceanic crust predominantly of basaltic composition, and a continental crust approximating to granite in composition. Surface geological processes operated within an atmosphere and a hydrosphere (oceans, lakes, etc. and groundwater), while internal geological processes were already being driven by convective processes not unlike those we recognize today. However, all these processes probably functioned on shorter time- and space-scales than they do today. The area of continental crust was much smaller than it is now, and that of oceanic crust correspondingly larger (oceans may have been more extensive, but there is no way of knowing how deep they were). The atmosphere was certainly very different – available evidence suggests it was dominated by nitrogen and carbon dioxide, with scarcely a trace of oxygen, as we noted earlier (see also Section 3.6.1). There is also evidence, as we shall see presently, that the Sun was some 25–30% weaker than it is today, so the amount of solar radiation reaching the Earth was correspondingly less. You should recognize that these are differences of detail and not of substance – though it must be admitted that a lack of oxygen would not be considered a detail by most life-forms. All of which prompts the question: at what point in the Earth's history can we place the first definite signs of life?

3.2 Geological evidence of the origin of life

The most reliable evidence for early life comes from fossils and from carbon isotope ratios which indicate photosynthesis (Box 1.4). The data suggest that life may have been flourishing 3.5 Ga ago, possibly 3.8 Ga ago, which in turn suggests that life may actually have *originated* by 4 Ga ago.

3.2.1 Stromatolites

In shallow coastal waters, characteristic mound-shaped structures called **stromatolites** (Figure 3.5) are built up by the accumulation of sediments consisting of thin gelatinous mats alternating with thin layers (laminae) of calcium carbonate. Organisms which form these mats, and precipitate the calcium carbonate, include the simple photosynthesizing nitrogen-fixing **cyanobacteria** or *blue–green algae* (also classified as Cyanophyceae). In cross-section, stromatolites have a layered structure like a stack of pancakes. Identical structures occur as fossils: the oldest stromatolites so far found are in sedimentary rocks 3.5 Ga old in northwestern Australia (Figure 3.5c).

Figure 3.5
(a) Modern stromatolites in Shark Bay, eastern Australia. The largest of the mounds is about 2 m across.
(b) Stromatolites about 2 Ga old from an iron-ore mine in Steeprock, Minnesota, USA.
(c) A cross-section showing the layered structure of a 3.5 Ga old stromatolite from the Warrawoona Group in the North Pole region of northwestern Australia. The sample shown is about 40 cm across.

The organisms which formed these ancient stromatolites may also have been cyanobacteria, because filaments and spherical structures in some ancient Australian stromatolites (Figure 3.6) look remarkably like modern cyanobacteria. However, this interpretation is still controversial, not least because some other modern photosynthetic bacteria (such as purple sulfur bacteria) also form mats or mounds. This is an important point with far-reaching implications for the environment of the early Earth. Cyanobacteria (but not some other forms of photosynthetic bacteria) carry out *oxygenic photosynthesis*, i.e. they release oxygen into the environment by means of the reaction you met earlier (Box 1.4):

$$nCO_2 + nH_2O \longrightarrow (CH_2O)_n + nO_2 \qquad \text{(Equation 1.4)*}$$

where $(CH_2O)_n$ represents a range of carbohydrate molecules (e.g. glucose, $C_6H_{12}O_6$, where $n = 6$), though fats and proteins are also produced in photosynthesis.

Cyanobacteria contain pigments that absorb ultraviolet wavelengths, and they have repair mechanisms that help them to resist damaging ultraviolet radiation. Sediment laminae between the gelatinous algal mats may also provide some protection. Similar unicellular organisms might thus have been well adapted to the shallows of ancient oceans, at a time when there was no shield of ozone in the upper atmosphere to absorb ultraviolet radiation from the Sun.

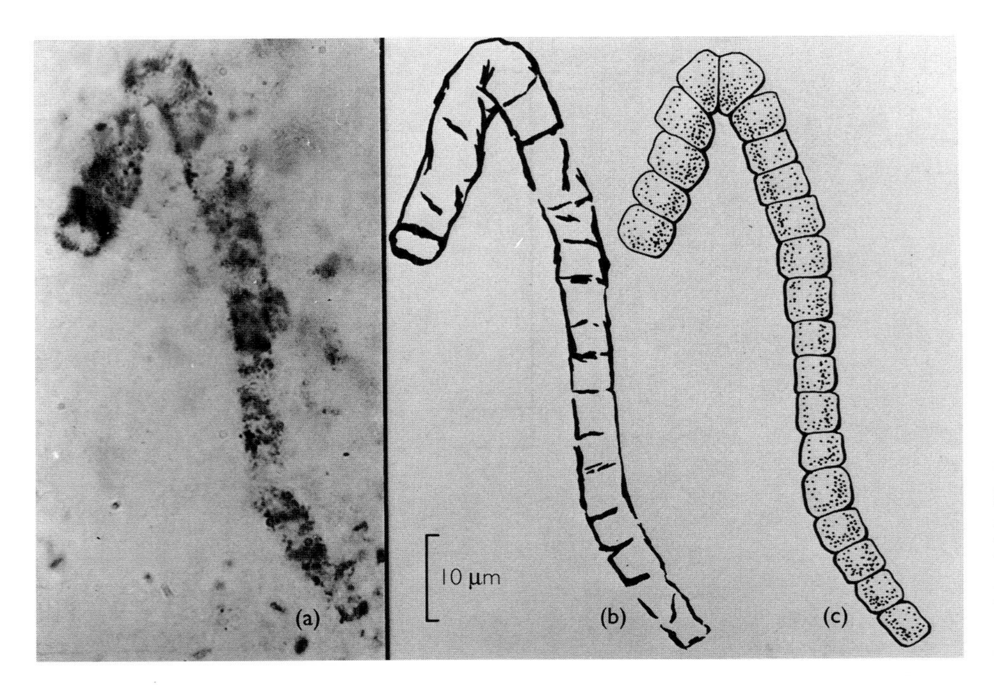

Figure 3.6
Segmented filamentous organic structures found in 3.5 Ga old fossil stromatolites resembling modern cyanobacteria.
(a) Photomicrograph, (b) line drawing, (c) reconstruction.

- Why was there no ozone shield at this time?
- Ozone (O_3) is formed from oxygen (O_2), and as the atmosphere was devoid of oxygen there could be no ozone.

However, no firm conclusion can be drawn from fossil evidence alone as to whether the dominant organisms 3.5 billion years ago were cyanobacteria with oxygen-producing ability or other non-oxygen-producing (*anoxygenic*) photosynthetic bacteria, of the kind known from modern oxygen-free (*anoxic*) environments, such as parts of the Black Sea.

Either way, all the evidence so far available strongly suggests that the type of metabolism dominant among organisms on the early Earth, 3.5 billion years ago, was **autotrophic** (self-feeding). In autotrophic metabolism, energy may be obtained from light (*photo-autotrophy* or *photosynthesis*, as in plants) or from simple, easily oxidized compounds such as sulfides (*chemo-autotrophy* or *chemosynthesis*, as in some bacteria), and building blocks are synthesized from molecules such as carbon dioxide and nitrogen (or nitrate or ammonium ions). A metabolism that obtains energy by consuming organic molecules is known as **heterotrophic** (other-feeding, as in all animals). It is possible that the very *earliest* form of life, 3.8 or even 4 billion years (or more) ago, was in fact heterotrophic. We shall explore this issue again in Chapter 4.

3.2.2 Carbon isotopes

Although biochemical remains are rare in ancient sediments, a portion of organic matter is often preserved as material called *kerogen* (from the Greek *keros*, meaning wax). Most ancient sediments have been heated by deep burial since they were deposited more than 3 billion years ago, and the kerogen in them is often chemically altered, which makes much of the structural chemical information it once contained indecipherable. However, it is possible to use the relative abundance of different carbon isotopes (Box 3.1) in ancient kerogen as a tool for unravelling the biochemistry of the early Earth.

Box 3.1 Carbon isotopes

Carbon consists of a mixture of two stable isotopes: ^{12}C (carbon-12) and the much rarer ^{13}C (carbon-13). In photosynthesis (Equation 1.4), fixation of the lighter $^{12}CO_2$ is favoured over that of the heavier $^{13}CO_2$, because $^{12}CO_2$ is taken up and diffuses into cells more rapidly and also reacts more readily. As a result of this isotope discrimination, or *fractionation*, organic matter produced by photosynthesis is enriched in ^{12}C and depleted in ^{13}C relative to the carbon in the inorganic carbon pool in the atmosphere and hydrosphere (mainly present as CO_2 gas, but also as carbonate and bicarbonate ions in solution). Enrichment or depletion of ^{13}C is expressed in terms of a **$\delta^{13}C$** value (δ is the lower-case Greek letter *delta*) which is calculated as follows. The $^{13}C/^{12}C$ ratio for the sample being investigated is given as a ratio to that of a carbonate standard in which the $^{13}C/^{12}C$ ratio is 1/88.99. The number one is subtracted from this value and the whole is multiplied by 1000 to give a $\delta^{13}C$ value in terms of parts per thousand or per mil (‰) of ^{13}C relative to the standard. This can be expressed in the following formula:

$$\delta^{13}C = \left[\frac{(^{13}C/^{12}C)\ \text{sample}}{(^{13}C/^{12}C)\ \text{standard}} - 1\right] \times 1000(‰)$$

(Equation 3.1)

- Using this formula, will the value of $\delta^{13}C$ be negative, positive or zero when the $^{13}C/^{12}C$ ratio is (a) equal for standard and sample, (b) greater in the sample, (c) lower in the sample?

- Because the number one is subtracted from the ratio of ratios, for (a) the value will be zero, for (b) it will be positive, and for (c) it will be negative.

The essential message you need to take from Box 3.1 is that when the sample is organic matter enriched in ^{12}C and depleted in ^{13}C relative to the standard, *the value of $\delta^{13}C$ will be negative*.

Figure 3.7a summarizes more than 10 000 measurements of carbon isotope ratios in sediments of all ages. The *mean* $\delta^{13}C$ value for 3.5 Ga old organic sediments (C_{org}) is about −26‰.

Figure 3.7b shows the range of $\delta^{13}C$ values for living (extant) autotrophs, present-day marine bicarbonate (in solution) and atmospheric CO_2. There is quite a range of values among autotrophs, reflecting their different styles of carbon-fixing reactions, but the $\delta^{13}C$ values are consistently negative and mostly between about −10 and −40‰.

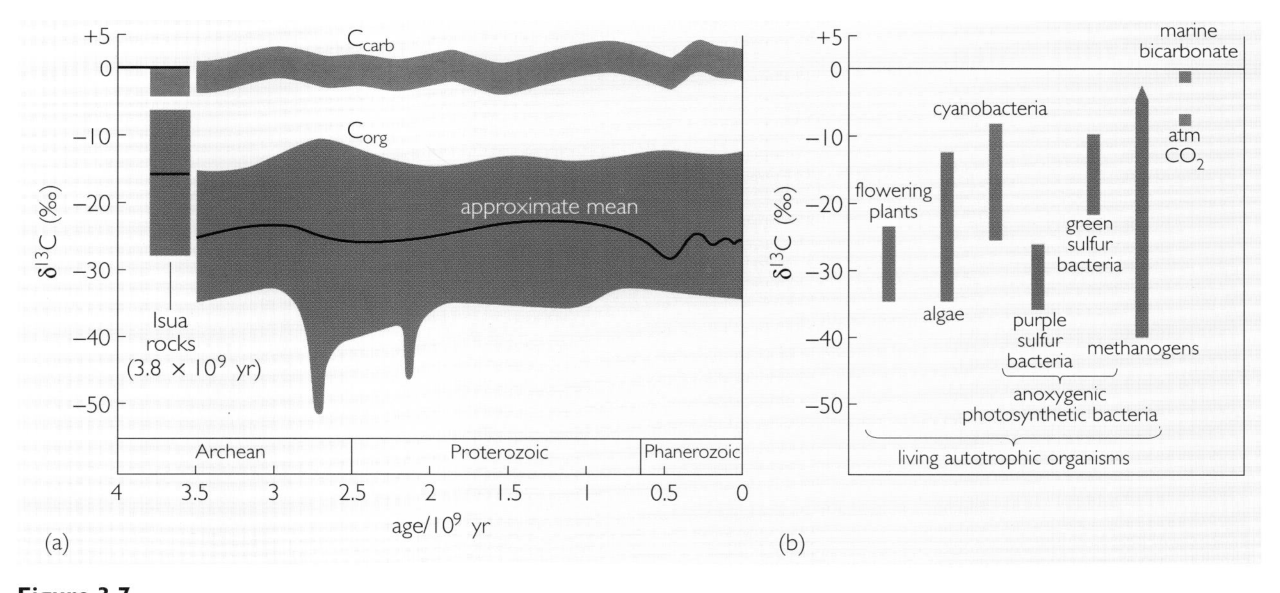

Figure 3.7
(a) Carbon isotope composition as $\delta^{13}C$ values for sedimentary carbonate (C_{carb}) and organic carbon (C_{org}) over 3.8 Ga of the Earth's history. Recent carbonate and marine sediments are represented on the right-hand side of part (a). (b) $\delta^{13}C$ values for present-day autotrophs and recently oxidized inorganic carbon. All organisms except green and purple sulfur bacteria and methanogens use oxygenic photosynthesis for carbon fixation. (Methanogens use hydrogen to reduce carbon dioxide in anaerobic environments.) The diagram is sometimes also known as a Schidlowski diagram, after its originator.

- What are the $\delta^{13}C$ values for the sources of carbon for photosynthesis (i.e. atmospheric carbon dioxide and marine bicarbonate, upper right of Figure 3.7b)?
- The values are in fact –7 and –1‰, respectively.

The range of isotopic ratios displayed in the C_{org} record in Figure 3.7a is similar to that observed in presently living autotrophs (Figure 3.7b), the majority of which fix carbon by oxygenic photosynthesis.

- Does this imply that oxygenic photosynthesis would have characterized organisms present on the early Earth?
- Not necessarily. Figure 3.7b shows that anoxygenic autotrophs (including methanogens) have $\delta^{13}C$ values in the same range as that of the C_{org} record in Figure 3.7a.

Although some scientists believe that there is evidence for oxygenic photosynthetic carbon fixation even in the oldest material examined – the Isua rocks from Greenland, formed 3.8 billion years ago (Section 3.1) which contain 0.6% of organic carbon – there is no consensus on this point, as you will see in Chapter 4. Figure 3.7a shows the Isua rocks have both *more* negative carbonate (C_{carb}) $\delta^{13}C$ and *less* negative organic (C_{org}) $\delta^{13}C$ values than those in later sediments. This could be because the rocks have been metamorphosed, i.e. subjected to high temperature and pressure, which is known to affect the carbon isotope ratios. The *original* $\delta^{13}C$ values for these rocks, before metamorphism, may have been the same as those in younger rocks. It is also possible that the different $\delta^{13}C$ values for C_{org} are real and were produced by organisms which fractionated carbon isotopes less strongly (cf. upper part of Figure 3.7a). Either way, autotrophs may well have been thriving 3.8 billion years ago. This conclusion would only be in error if there were some other global process capable of mimicking the isotope fractionations in photosynthesis.

Question 3.2

Fossil-bearing sediments recovered from the 2.6 Ga old Fortescue formation in Australia contain around 1–2% carbon by mass. When this carbon was analysed for its isotopic composition, the $\delta^{13}C$ values obtained averaged around –50‰. According to the data in Figure 3.7b, which, if any, of the groups of organism listed might have been the major contributor to the organic carbon in these sediments?

Figure 3.7a, in fact, shows a pronounced negative 'blip' at around 2.7 billion years, with carbon isotope values down to about –50‰ (as in Question 3.2). Just why this occurs, and why there is another blip at around 2.1 billion years, is not known. It could be a consequence of further isotope fractionation resulting from the activities of methanogenic bacteria (methanogens, Figure 3.7b) on the organic matter (precursor of kerogen) in the sediments.

So what did the surface of the Earth look like when life first appeared, certainly by 3.5 Ga ago, possibly 3.8 Ga ago, perhaps even before that? The oldest rocks give some indication of surface conditions around 4 billion years ago, but as we saw in Chapter 2 the Earth formed 4.6 billion years ago – so there is a large hiatus of about 600 million years between the birth of the Earth and the preservation of the oldest rocks. It is a very large time interval. What happened during those first 600 million years? The Earth itself contains no record of the period between 4.6 and 4 billion years ago, so we must look outside the Earth for our evidence.

Figure 3.8
Michelle Knapp's car which was damaged when it was hit by a 12 kg meteorite.

3.3 Earth's cratering record

At 8pm on 9 October 1992, Michelle Knapp of Peekskill, New York State, USA was dismayed to hear the crunch of buckling metal. Her 1980 Chevrolet Malibu, parked outside her home, had been hit by a 12 kg meteorite (Figure 3.8). It was one of about 70 fragments of a body observed as a green–white fireball moving along a 700 km path from the direction of Kentucky. Analysis of video footage showed that the 'shooting star' entered the Earth's atmosphere at an angle of 3.4° to the horizontal. It originated from the asteroid belt beyond the orbit of Mars (Figure 2.7). Apart from its unusually low entry angle, the Peekskill meteorite was unexceptional. Michelle was lucky: atmospheric drag had slowed the projectile to a sedate 0.04 km s^{-1} (150 km h^{-1}) from its entry speed of 20 km s^{-1} (75 000 km h^{-1}). Otherwise her car and her home could have been vaporized by the impact. The car is now a collectors' item.

Michelle Knapp's experience reminds us that the Earth is still being bombarded by extraterrestrial bodies, and you will recall from Chapter 2 that these include meteorites, small asteroids, cosmic dust and the occasional comet.

Question 3.3
Why do comets reach the Earth's surface less frequently than meteorites?

The Earth's surface carries a surprisingly large number of scars left by the impacts of asteroids and large meteorites, and it is instructive to consider the energy that is produced when these objects hit the Earth. As an example, we'll calculate (in Question 3.4) the energy that would have been released had an asteroid seen in 1989 hit the Earth – luckily it turned out to be a 'near miss' and passed by the Earth no nearer than the Moon.

(a)

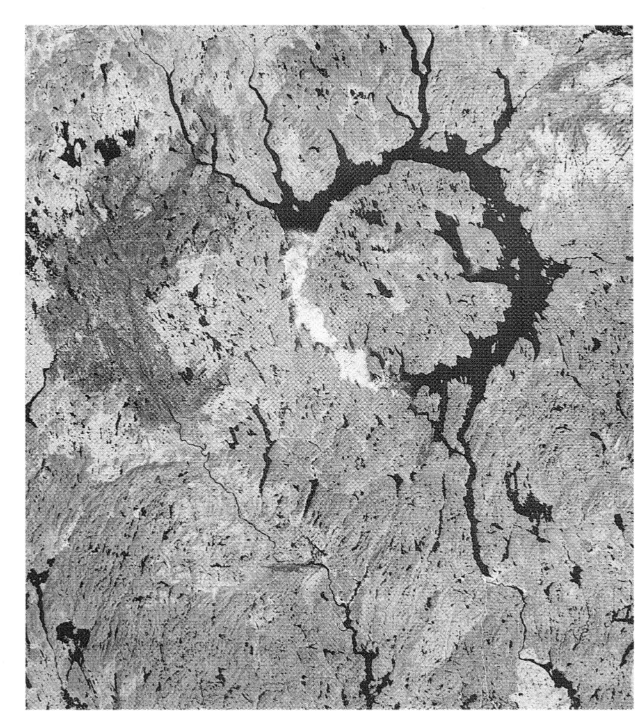

(b)

Figure 3.9
(a) The celebrated Barringer Meteor Crater in Arizona is about 1 km across, 200 m deep and has a rim 50–70 m above the level of the surrounding plains. It was formed 50 000 years ago.
(b) A satellite picture shows a circular lake at Manicouagan in Quebec, Canada, outlining the boundaries of an impact structure formed 210 Ma ago. It is about 65 km across.

Question 3.4

In March 1989, an asteroid 0.5 km in diameter passed within half a million kilometres (5×10^5 km) of the Earth, travelling at a speed of around 20 km s^{-1}. If we assume that the asteroid had a spherical shape and a density of 3000 kg m^{-3}, what was its kinetic energy?

Begin by using the formula $\frac{4}{3}\pi r^3$ to calculate the asteroid's volume (remember the radius, r, is half the diameter and use $\pi = 3.14$). Then multiply the volume by the density given above to calculate the mass of the asteroid. Then use the formula $\frac{1}{2}mv^2$ to calculate the kinetic energy of the asteroid's motion, where m is its mass (in kg), which you have just calculated, and v is the speed, which you must convert to m s^{-1} (1 km = 10^3 m).

Your answer will be in kg m^2 s^{-2}, the basic units of energy, for which (cf. Question 2.1) the SI unit is the *joule*, J.

The answer to Question 3.4 is a very large number which has little meaning by itself. To grasp its significance, imagine that the asteroid had hit the Earth, and compare the energy it would have produced on impact with the values given in Table 3.1.

Table 3.1 Comparisons of energy produced by various processes (see also the notes below).

Phenomenon	Energy produced in a second/J s^{-1}
simultaneous detonation of the total global nuclear arsenal	8×10^{19}
impact of a 0.5 km asteroid (as calculated in Question 3.4)	3.9×10^{19}
solar heating of the Earth	1.5×10^{17}
Earth's internal heat production by radioactive decay	1.7×10^{13}
energy used by human societies today	1.3×10^{13}

Notes to Table 3.1

1 The SI unit of energy is the **joule**, and as noted in Question 3.4, this can be expressed in the basic SI units of kg m^2 s^{-2}. Energy is defined as the capacity to do work, but for our purposes we can recognize that energy manifests itself in a number of ways: heat energy, mechanical energy, the kinetic energy inherent in motion and the potential energy of, for example, any object above ground level that could fall towards it.

2 *Power* is the rate of expenditure of energy, or rate of doing work, and so has units of joules per second, J s^{-1}, more commonly known as the **watt**, W (which is the SI unit of power). Therefore the values given in Table 3.1 could have been expressed in watts.

3 The energy produced by the impact of the asteroid described in Question 3.4 would have been colossal but would still have been only about half that which would result from instantaneously setting off the global nuclear arsenal (a sobering thought indeed, for those studying the subject of Earth and life). Both these awesome figures are six orders of magnitude greater than the 'artificial' energy we humans expend globally each second for transport and for domestic and industrial purposes – the amount of energy we produce by our metabolic (biological) processes is paltry by comparison, some two orders of magnitude less than the smallest value in Table 3.1. Our energy use has now grown to rival the energy produced by radioactive heating in the Earth's interior, but both are four orders of magnitude less than the energy which the Earth as a whole receives from the Sun each second.

You may have wondered what size of crater the asteroid mentioned in Question 3.4 would have made had it hit the Earth. For comparison purposes, it is estimated that some 10^{24} J were produced by the asteroid which formed the Manicouagan crater (Figure 3.9b) and that it may have had a mass of 10^{11}–10^{12} tonnes (i.e. 10^{14}–10^{15} kg), corresponding to a diameter of between 1.5 and 2 km, i.e. three to four times the size of the asteroid described in Question 3.4. By contrast, the small asteroid which formed the Barringer Meteor Crater (Figure 3.9a) is estimated to have been about 60 m in diameter.

The Manicouagan impact happened 210 Ma ago. Have we any grounds for thinking that such an event could or could not happen during our lifetime? The answer is that we cannot be sure either way, but modern technology might at least give us some warning of any impending impact – though it would probably not do us much good, as even on the side of the Earth away from the impact, the effects would still be catastrophic.

■ If we look back at Figure 2.15, which illustrates the favoured theory of the formation of the Earth by aggregation and accretion, can we say that the likelihood of such impacts has decreased with time, so they are less likely now than in the geological past?

■ Looking at Figure 2.15, it is not hard to envisage that there must have been many more collisions early in Earth's history than now.

On balance then, we might expect that the incidence of such events did indeed decline with time, as larger bodies in the Solar System 'mopped up' or 'swept up' smaller ones, mainly through accretion, or more rarely through capture of larger bodies as planetary satellites. So you could find the information given in Figure 3.10 to be somewhat counter-intuitive, in that it seems to contradict this general conclusion.

Question 3.5
What does Figure 3.10b suggest at first sight about how the frequency of impacts has changed with time? Can you account for the apparent conflict with expectations, in the light of what you read earlier in this chapter?

Craters on the sea-bed must also have been formed by impacts in the oceans. However, they would not survive for long, partly because of burial by sediments accumulating on the sea-bed, and partly because of internal convection (Figure 3.4) and the plate tectonic cycle. There is no oceanic crust older than about 160 Ma – all oceanic crust older than this has long since been subducted back into the mantle (Figure 3.3).

All in all, then, we could surmise that impacts were probably at a maximum during and just after formation of the Earth, i.e. early in the 600 Ma hiatus between planetary formation and preservation of the oldest rocks (Section 3.2.2) – but we shall see presently that this conjecture may turn out to be invalid.

■ Should we expect the Earth to have had a layered structure like that in Figure 3.2 from the start?

■ Not really, if the Earth was formed (Figure 2.15) from what must have been essentially a random accretion of a variety of bodies, including comets, asteroids and dust grains, comprising various combinations of metal, sulfide, oxide and silicate along with complex organic molecules, condensed gases and ice.

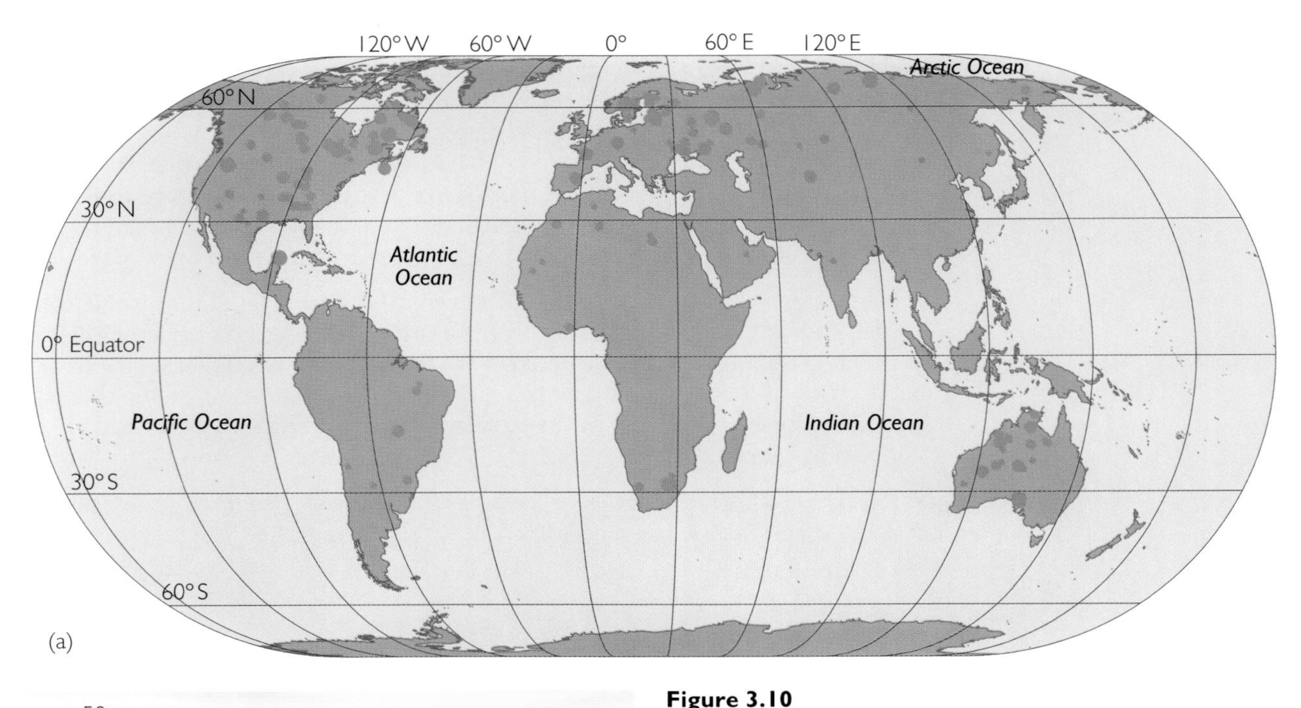

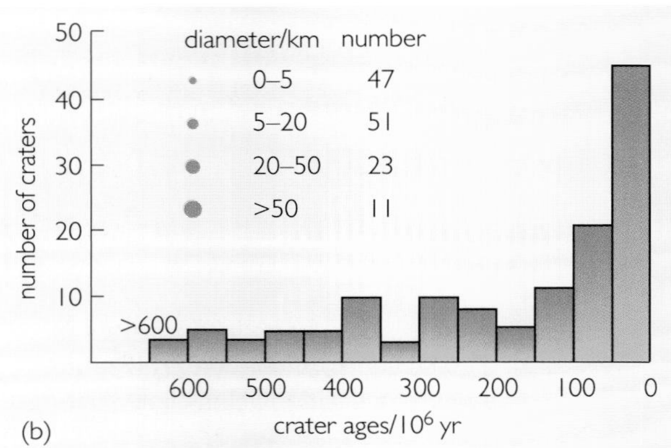

Figure 3.10
(a) Location of known impact craters (1992 data). Data are biased towards North America and Australia, partly because these continents contain the largest areas of ancient continental crust, and partly because they are regions where systematic searches have been started. Other impact sites have been found since 1992 but they do not alter the basic pattern shown here. The key for the crater size is given in part (b).
(b) Histogram showing numbers of craters shown in part (a) plotted against inferred age.

A possible analogy for the initial result of such aggregation would be something like a fruit-cake mix. However, modelling of the process suggests that the temperatures reached during accretion (due to kinetic energy being converted into heat energy) would have been high enough to cause melting and homogenization of the different components. Decomposition of unstable compounds and the escape of volatile compounds (gases and vapours) would have contributed to the gaseous envelope (the primitive atmosphere) that developed round the young planet soon after it formed. Generally similar scenarios would presumably have applied to the other planets, too.

- Have we any information about the overall average bulk composition of this early homogeneous Earth?
- We have only the indirect evidence provided, for example, by meteorites and by stellar (including solar) spectra (Section 2.3.2) and inferences such as those upon which Figure 3.2 is based.

The favoured hypothesis for the average bulk composition of the early Earth is that it approximated to that of carbonaceous chondrites (Figure 2.10). So, if you were to plot average elemental abundance data for the *whole* Earth against data for the Sun, they would lie close to the straight line in Figure 2.11.

- If you look back at Figure 2.11, you will see that oxygen plots in the top right of the graph. How would you reconcile this with the statement we made earlier that there was probably no oxygen in the atmosphere of the early Earth?
- All the Earth's oxygen was initially combined in silicates and oxides, apart from a little combined in water and complex organic molecules.

It is crucial to realize that even the present-day amount of free oxygen in the atmosphere is *minute* when compared with that combined in silicate rocks.

There can have been no more than trace amounts of free oxygen in the atmosphere until there was enough oxygen-producing life available to generate more (a point made in Chapter 1, see also Chapter 4). In fact, as outlined in Section 2.1, small amounts of oxygen *are* produced in the atmosphere by photolytic dissociation of water vapour by the action of ultraviolet solar radiation:

$$2H_2O + \text{energy} \longrightarrow 2H_2 + O_2 \qquad \text{(Equation 3.2)}$$

Oxygen, being a highly reactive gas, would soon be used up in oxidation reactions at the Earth's surface or in the atmosphere, while hydrogen, being the lightest of all the elements, would tend to escape from the atmosphere to space.

Question 3.6
Should we expect the average bulk composition of the present-day Earth (Figure 3.2) to be similar to that of carbonaceous chondrites?

Unfortunately, none of this helps us to determine at what stage in the 600 million years following accretion the Earth became differentiated into its layers (Figure 3.2), nor does it tell us about conditions on the Earth's surface during that time. All we know so far is that the incidence of impacts probably decreased somewhat with time. But that is all.

The Moon, however, is pockmarked with craters, some of them very large, as you can see even with a pair of binoculars (Figure 3.11). Interestingly, it seems that no new craters have been observed on the Moon since detailed observations of the lunar surface began (i.e. in the last couple of hundred years or so that telescopes have been available). That doesn't mean that the Moon has been immune from impacts during that time; it means only that there have been no impacts large enough to make a crater that could be seen from Earth. We know that the Moon has no atmosphere, as it is too small for its gravitational field to be strong enough to prevent gas molecules from escaping; and there can be no weathering and erosion on the lunar surface. Nor is there any evidence of internal activity of the kind that characterizes the Earth. Perhaps, then, the Moon's myriad craters may preserve a long record of meteorite impacts, from which we might glean some insights into the early history of Earth. Let us see.

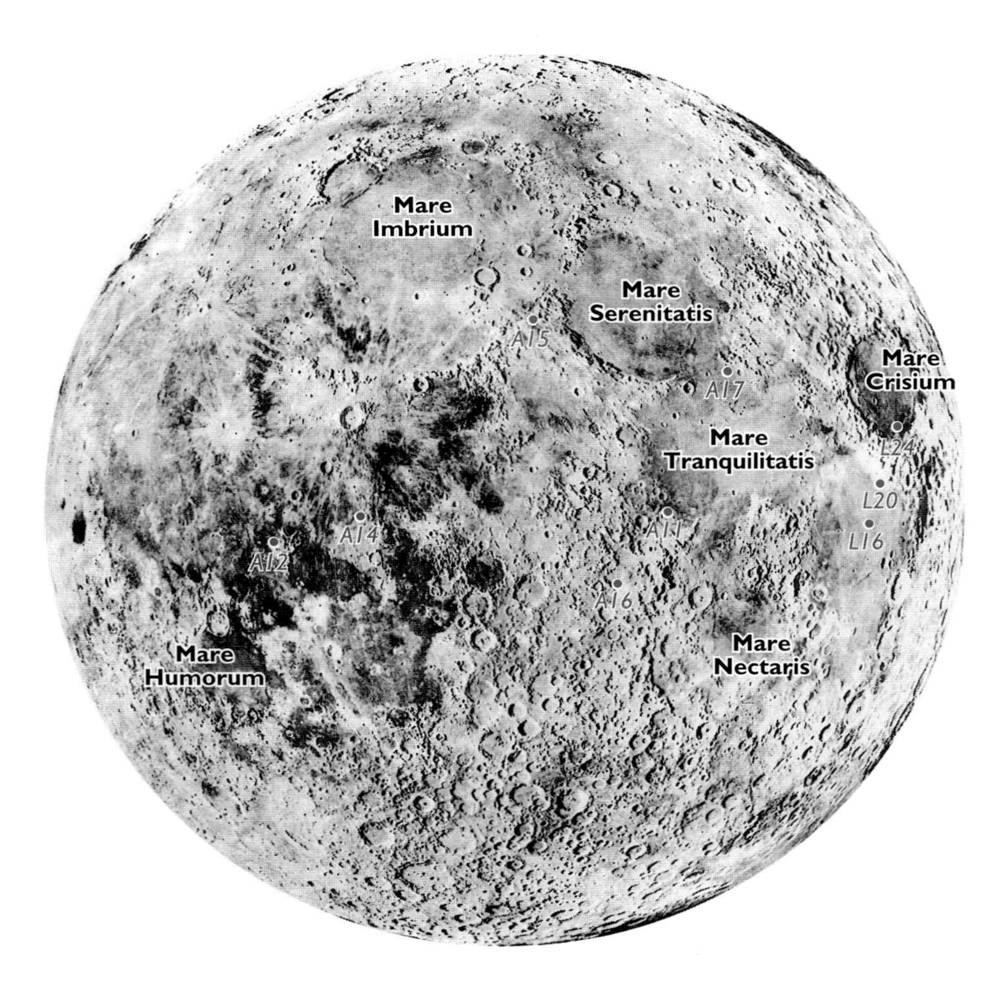

Figure 3.11 Photograph of the nearside of the Moon showing the dominance of circular features formed by impacts. The labels refer to the landing sites of the Apollo missions, *A* (USA) and Luna missions, *L* (USSR) to the Moon. (The Moon's cratering record is discussed in Section 3.4.1.)

3.4 The Moon

An understanding of the Moon is important because it is our closest neighbour in space and so must have been formed of material similar to that making up the Earth. The Apollo space missions of the 1970s confirmed earlier inferences that it is a geologically inactive body. The Earth's surface provides a record of continuous internal and surface processes, while that of the Moon records mostly the results of bombardment since its formation (Figure 3.11), and it is thus a source of information about *external* effects on planets dating from that time.

The Moon's average density of 3340 kg m^{-3} is less than that of the Earth (5520 kg m^{-3}), which means it must be deficient in some heavy element(s) compared with the Earth. Only iron is sufficiently dense and abundant in the cosmos to explain this difference, and there is a variety of evidence which shows that the Moon either has no iron-rich core or, at best, only a very small one. Estimates of the bulk composition of the Moon suggest that it is similar to that of the Earth's mantle (Figure 3.2), i.e. it is depleted in iron.

Question 3.7

(a) Where is most of the Earth's complement of iron now to be found?

(b) Why can't we appeal to a heating event early in Solar System history to explain the deficiency of iron in the Moon?

Comparison of the lunar materials (brought back by the Apollo missions) with the Solar System chemical 'standard', the carbonaceous chondrites (Figure 2.11), showed that the Moon is depleted in volatile elements. This depletion is more severe than occurs in the Earth, as we might expect from the smaller size and hence lower gravity of the Moon. Lunar rocks are formed of silicate minerals almost identical to those found on Earth, and they are all either igneous rocks, or shattered debris cemented by glass (the product of rocks melted by impact heating). The Moon has no sedimentary rocks of the kind formed by the action of flowing water.

The Apollo samples were dated radiometrically (i.e. using the known decay rates of radioactive elements in the samples). The oldest samples, around 4.4 to 4.5×10^9 years old, all came from the paler areas on the Moon (Figure 3.11) which are the areas with the most rugged topography. These regions of the Moon are known as the lunar Highlands, and they look paler because the rocks consist almost entirely of a white calcium aluminium silicate known as *plagioclase feldspar*, which is well known to geologists on Earth as the principal constituent of basaltic rocks. Plagioclase feldspar has a low density compared with that of other lunar materials, and so the paler Highland rocks have a lower density than the rocks that underlie the darker maria ('seas', singular *mare*, plural *maria*) on the lunar surface. (These are not 'seas' in the terrestrial sense.) It is this density difference that results in the Highland areas standing about 2.75 km higher than the maria. This relationship, together with variations of the lunar gravitational field, indicates that the feldspar-rich Highland rocks must be around 20 km thick. Rocks of the darker, denser maria are similar to basalt lavas that are common on the Earth, except that those on Earth contain up to 2% by mass of water chemically combined within the minerals of which they are composed – whereas the lunar 'basalts' contain no water. Apollo maria samples have been dated at between 3.8 and 3.2×10^9 years old.

- Does this description of the distribution of the two main kinds of lunar rocks have any analogy with the situation on Earth?

- We know that Earth's granitic crust is both less dense and thicker than its basaltic crust; the granitic crust forms 'higher ground' (the continents), whereas basaltic crust forms 'lower ground' (the ocean basins). However, while the lunar maria are basaltic, the lunar Highlands are *not* granitic. (Nor, it must be emphasized, are they basaltic, for although plagioclase feldspar is indeed the principal constituent of basalts, these rocks contain other minerals as well.)

Formation of the 20-km-thick feldspar-rich 'crust' of the lunar Highlands is believed to have been achieved by the low-density feldspar floating upwards as it crystallized from a 200-km-deep layer of completely molten lunar mantle (which is thought to be generally similar in composition to the Earth's mantle, Figure 3.2). This implies that the Moon once had a deep, all-enveloping *magma ocean*.

Question 3.8

Why must this have occurred early in the Moon's history?

The present consensus about the origin of the Moon is that a planet about the size of Mars made a glancing collision with the early Earth. There would have been a cataclysm of more than Biblical proportions (Figure 3.12), with part of the Earth's mantle being torn off, vaporized and flung into orbit as a dense, incandescent cloud. Cooling and accretion of this debris then formed the Moon, with the gravitational energy lost in the accretion being transformed into heat which melted much of the Moon and so formed the magma ocean. The immense temperatures involved led to the loss of volatile material from both the Moon and the Earth, with proportionately more being lost from the vaporized 'Moon matter'. In other words, the collision would have blasted away any atmosphere and hydrosphere that there might have been on the early Earth. As the event occurred soon after formation of the Solar System 4.6 $\times 10^9$ years ago, we have no means of knowing what that atmosphere was like, still less whether the surface of the Earth was cool enough to support a hydrosphere of liquid water.

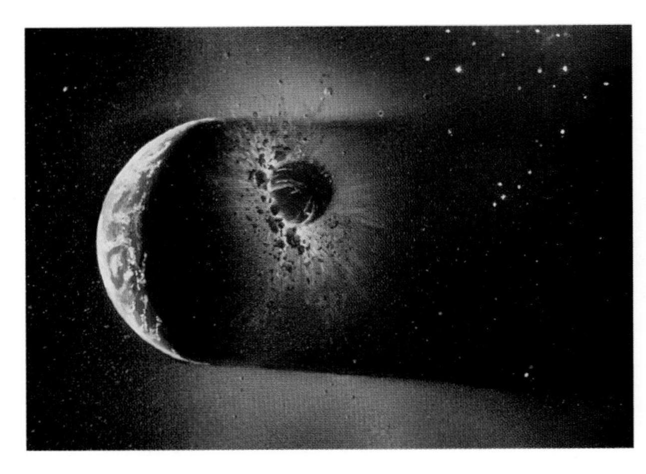

Figure 3.12
Artist's impression of the collision process that may have formed the Moon soon after the formation of the Solar System about 4.6 Ga ago.

- Bearing in mind the relative densities of the Earth and the Moon, would you expect this collision event to have occurred after or before the Earth had acquired its metallic (iron–nickel) core?
- The Moon's low density and inferred iron depletion suggest that it happened *after* the Earth had acquired its own separate core and mantle.

If the Earth's core had already formed by 4.4 to 4.5 $\times 10^9$ years ago (the age of the Moon's Highland rocks, formed from its magma ocean), then the Earth's mantle must have begun to approach its present composition, although at that time there may well have been no crust.

Radiometrically determined ages of lunar rocks fall within the 600 Ma 'hiatus' between the formation of planet Earth and the preservation of the oldest terrestrial rocks. The very ancient *lunar* rocks might therefore enable us to infer something about conditions on the early Earth at that time. Before moving on, however, you should realize that the radioactive 'geological clocks' used to measure the ages of rocks are 'reset' only by heating the rocks to several hundred degrees Celsius. This means, for instance, that lunar Highland rocks with ages of 4.5 billion years have remained unaffected by any heating event virtually since the Highlands themselves were formed.

- According to the age range of lunar rocks you read about earlier, when was probably the last time that the Moon experienced a significant heating event?
- Presently available evidence suggests that the last significant heating event to affect any part of the lunar surface was 3.2 billion years ago, the age of the youngest maria rocks so far found.

3.4.1 The story in the lunar craters

Because of its geological 'lifelessness', the Moon preserves a record of what has happened in near-Earth space during the 600 Ma or so that elapsed between the formation of the Earth and the preservation of the oldest terrestrial rocks (Section 3.1.1). It provides a sort of 'witness plate' for the history of the early Earth.

A photograph of the nearside of the Moon (Figure 3.11) is a familiar image. The 'Man in the Moon' is picked out by the dark maria and the light Highland surfaces. Figure 3.13 reveals more details of these two types of surface. The Highlands are rugged, with many shadows, whereas the mare has a flat surface on which the only

Figure 3.13
Oblique view of part of the Moon (from Apollo 17), showing Highland material to the left and darker mare material to the right.

shadowed relief is due to elongated rilles (trenches), the result of later fracturing (faulting) of the lunar crust. Both surfaces are pockmarked with impact craters. On this image, the Highlands appear to be more 'poxy' than the mare, though the craters seem to be sharper on the mare surface.

The differences between the Highlands and the maria enable us to reconstruct the Moon's bombardment history. The Highlands preserve a record of impacts between about 4.5 and 4.4 billion years ago in their radiometric ages, whereas the ages of the maria rocks provide a record covering the period between 3.8 and 3.2 billion years ago. We can make a rough time division between what happened *only* on the Highland surface, the early record, and a record of later events that affected both surfaces but which are most clearly preserved on the maria. Figure 3.14 shows higher resolution images of parts of a Highland area and a mare surface.

Question 3.9
Which of the two areas shown in Figure 3.14 has (a) the greater number of craters, and (b) the larger craters?

Given the vast amounts of energy produced by impacts (Question 3.4), you might wonder how the lunar Highland rocks seem to have preserved their very old ages of 4.4 to 4.5 billion years, despite the intensity of bombardment manifest in Figure 3.14. We can only suggest that the sampled rocks came from sites *between* impact craters and so experienced only moderate heating, insufficient to reset their radioactive 'clocks'. If that is a valid inference, and your answers to Question 3.9 can be applied to the Moon as a whole, then there were more relatively large impacts per unit time before maria formation; and the number and size of all projectiles per unit time decreased markedly after the maria formed.

(a)

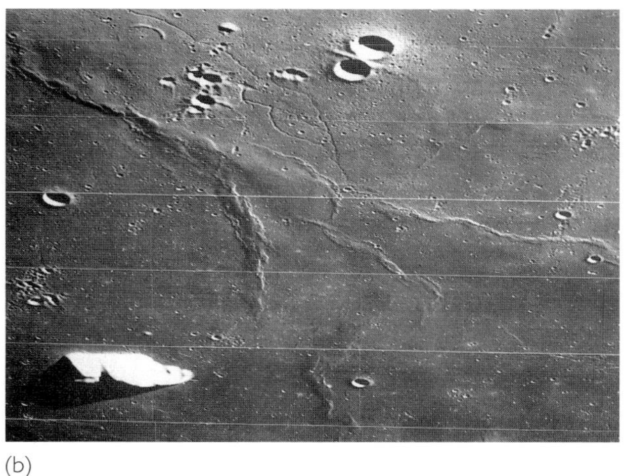

(b)

Figure 3.14
(a) A region of the lunar Highlands, about 300 km across; (b) a mare surface at the same scale, but viewed obliquely.

If you allow for the curvature of the lunar surface, then Mare Imbrium, Mare Serenitatis and Mare Crisium (Figure 3.11) appear to be large circular structures (some more than 1000 km across). They are generally regarded as giant ringed impact basins that have been more or less filled by lavas welling up from depth. The lavas were formed by partial melting of the underlying lunar mantle rocks heated by the energy generated by the huge impacts. In places, the lavas overflowed the margins of the basins to overlie densely cratered Highland rocks. So the age of the impact basins is somewhere between the two major components of lunar geology, i.e. between 4.4 and 4.5 billion years (the Highlands) and 3.2 billion years (the youngest maria). Apollo samples from the Highland landing sites provided materials to date these giant events. The materials used for dating were glasses formed by the chilling of rocks melted and flung out during the ringed-basin impacts (like the spatter of droplets you see when you throw a stone into a pond). Their ages all fall within a 200 Ma period between 4.0 and 3.8 billion years ago. No glasses older than that were found during the Apollo missions, suggesting that very large (perhaps the largest) impact events occurred between about 400 and 700 million years *after* the Moon acquired a solid crust.

In other words, very large projectiles were hitting the Moon quite a long time after it had formed, and *not* soon after its 'escape' from Earth, as we might have supposed from Figure 2.15 and related text. The number of large bodies whizzing about the Solar System would be expected to have been at a maximum round about 4.6 to 4.5 billion years ago. However, if the inference drawn from the ages obtained for lunar samples is valid, very large impacts were still occurring long after formation of the Moon, and not merely in the period soon after its formation (and that of the Earth).

■ What does this imply about the bombardment history of the early Earth?

■ The Earth would also have been subjected to impacts by very large bodies during the same time interval, i.e. between about 4.0 to 3.8 billion years ago.

Moreover, the Earth is a denser body than the Moon and it has a stronger gravitational field, so it would attract more (and larger) meteorites and asteroids; so bombardment might have been more severe than on the Moon.

An intriguing aspect of this story is that the ages of the impact glasses on the Moon seem to be nowhere younger than 3.8 billion years, whereas basaltic rocks filling the maria are as young as 3.2 billion years. As the maria basalts were generated by the partial melting caused by the heat energy released in the giant impacts, it might seem surprising that several hundred million years could elapse between impacts and the generation, eruption and cooling of the resulting lavas. The explanation may have to do with the absence of convection within the Moon, so heat transfer could occur only by the much slower process of conduction (Section 3.1.2). It is also possible that mare basalt eruptions occurred over a long period of time, and that the 3.2 billion year ages represent merely the latest (youngest) of these eruptions.

3.5 A glimpse of the early Earth

Whatever the early Earth may have looked like before about 3.8 billion years ago, it is hardly surprising, in the light of the discussion so far, that so few traces of its early history are preserved in the oldest rocks. Other evidence was obliterated by a combination of cataclysmic meteorite impacts and vigorous convection in the mantle.

- What about liquid water? Could this blasted early Earth have sustained a hydrosphere?
- We can only speculate that while there might have been periods when surface temperatures were below about 100 °C, enabling water to condense, it is at least possible that the energy liberated in large impact events – especially those in the 4.0 to 3.8 billion year interval – could have kept surface temperatures above the boiling temperature of water.

We can also speculate that the outermost parts of the solid Earth were generally of basaltic composition, forming the floors of any ocean basins that may have developed if surface temperatures fell sufficiently for water to condense. At such times, land areas would have been very limited in extent, and probably confined chiefly to volcanoes associated with vigorous mantle convection. It is at least plausible to suggest that the Earth's interior was too hot and convection accordingly too rapid to allow formation of a granitic crust until towards the end of the heavy bombardment period (4.0 to 3.8 billion years ago).

What about life on this early Earth? Early life-forms would probably have needed liquid water to survive, though the water could have been as hot as 80 °C or more. Modern hot springs contain communities of **thermophilic bacteria** thriving at 50–60 °C, and there are varieties which can tolerate considerably higher temperatures than that (see Section 4.4.2). Still more amazing, in the early 1990s came news that bacterial communities had been found living in oil reservoirs thousands of metres below the surface, at temperatures well in excess of 100 °C and pressures of 450 atm or more (i.e. about 10^7 Pa). These subterranean bacteria require neither sunlight nor oxygen (they are *anaerobic*), and obtain their energy by oxidation of organic compounds in the petroleum, via reduction of sulfate (SO_4^{2-}) ions to sulfides (e.g. H_2S). They probably evolved from thermophilic surface-dwelling anaerobic bacteria (because they 'feed on' organic matter that originally formed at the surface). This discovery raised the possibility that the search for present or past life on Mars (Section 1.1) did not go deep enough – a topic we shall return to in Chapter 4.

At the other end of the temperature spectrum, however, there are micro-organisms living in cracks in rocks in the dry valleys of present-day Antarctica, suggesting that cold dry conditions cannot be entirely ruled out as a possible environment for life on the early Earth (see Section 3.6.2, item 3).

Life-forms may have become established on Earth more than once during the 600 Ma hiatus after its formation (and in more than one possible environment), but they could have been obliterated when major impacts occurred. It is estimated that during the period of maria formation (4.0 to 3.8 billion years ago), the largest impacts on Earth were capable of forming craters up to 5000 km across. It is hard to see how any life-form could have survived the global consequences of such impacts.

Which brings us back to the oldest rocks, with their traces of possible organic remains (Section 3.2.2). If these traces really are from fossil organisms, we can be confident that life on Earth had already begun when the sediments containing them were deposited, and that the Earth may by then have had a permanent hydrosphere – so helping to ensure the future of that life.

Question 3.10

But can we infer from the lunar record that life at that point was immune from further dangers of obliteration by asteroid and meteorite bombardment?

And what of the origin of that life? How and where on Earth did it begin? Or did it perhaps not begin on Earth at all, but arrived here from elsewhere, perhaps repeatedly, only surviving to develop and begin to evolve into the enormous variety of organisms we see today, when the Earth had oceans to support it? We address these issues in Chapter 4, but let us now see if we can conjure up a picture of the conditions under which those earliest life-forms may have existed. Before we do that, attempt Activity 3.1 by way of revision.

Activity 3.1

Examine Figure 3.15 and in the light of what you have read so far in this chapter and in Chapter 2, devise a caption for the picture. Include in your caption some indication of the likely period of time represented in each component of the picture, and for part (e) specify the nature of the 'metals' and the 'light silicates'. What scale(s) would you put on parts (d) and (e) of the picture?

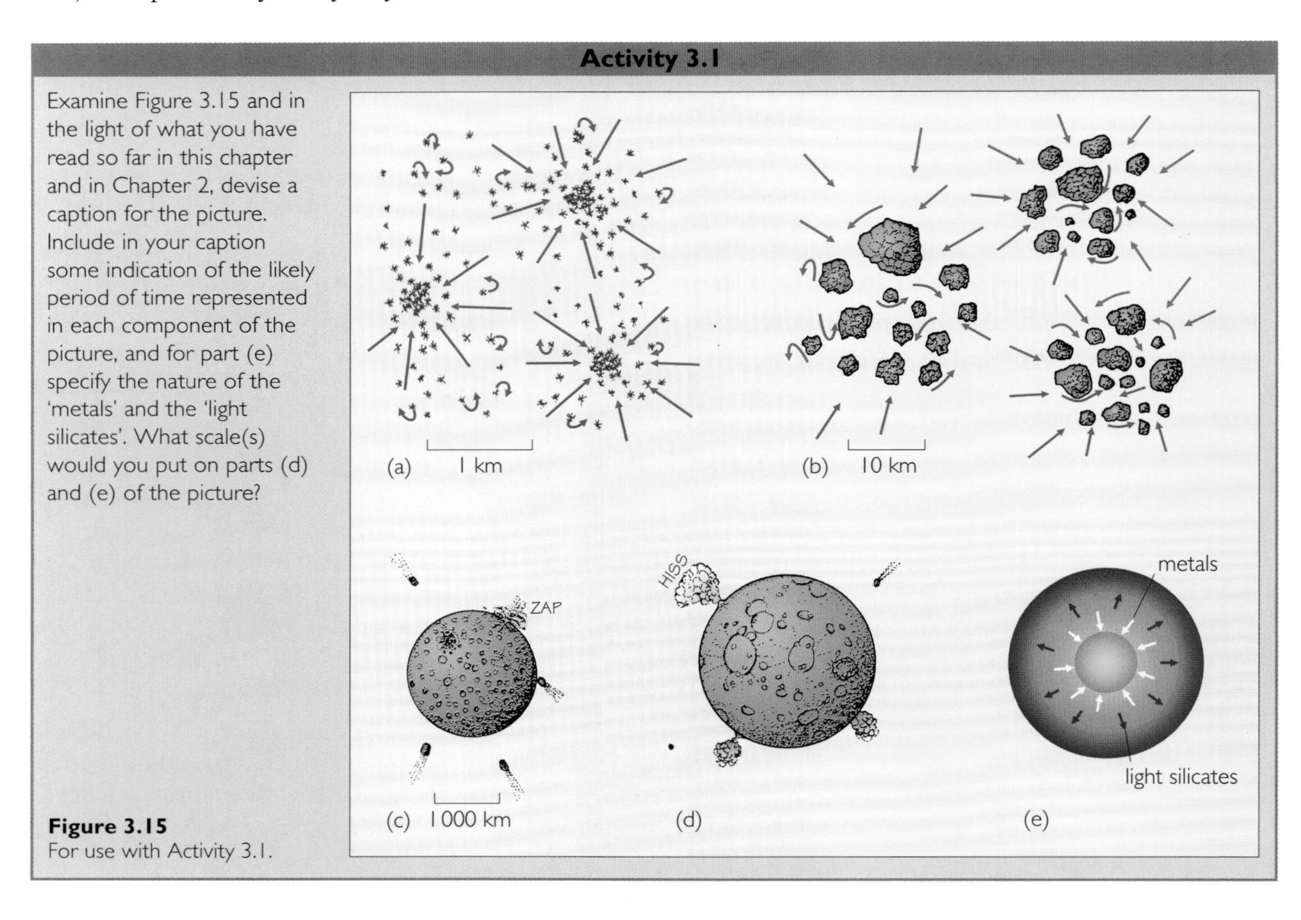

Figure 3.15 For use with Activity 3.1.

3.6 The Hadean Earth

The 'geological' period before the oldest rocks, i.e. from about 4.6 to 3.8 billion years, is called the **Hadean** (from the Greek for the Underworld), partly perhaps because it is widely believed that conditions then were fairly hellish. The Hadean is only very sparsely represented in the geological record, but it would be useful to see if we can get some idea of what the early Earth looked like at least at the end of this period, because it seems likely that this was also the 'dawn of life'.

3.6.1 Early atmosphere and hydrosphere

Large amounts of volatile materials (see, for example, the molecules listed in Tables 2.1 and 2.2) became incorporated into the Earth as it formed. However, the heat generated during accretion (Section 2.4.4), as well as during formation of the Earth's core and of the Moon (Section 3.4), would have driven off most of these original volatile materials. Those which remained within the Earth have been escaping from its interior (**outgassing**) ever since. Outgassing rates must have been much greater during the Hadean than they are now, because rates of internal convection (Figure 3.4) were greater. Most of the atmosphere and hydrosphere probably originated from within the Earth during the Hadean, after formation of the core and 'birth' of the Moon. The hydrosphere may subsequently have been augmented by water from cometary impacts (indeed, some scientists believe most of the world's water is from this source).

Various lines of evidence, including the composition of gases emitted from modern volcanoes, our knowledge of the present composition of the atmosphere, and observations of conditions on other planets, have led to the conclusion that after formation of the core, the Hadean atmosphere became dominated by nitrogen (N_2), carbon dioxide (CO_2), possibly sulfur oxides (especially SO_2), and water vapour (as there were oceans present). It was devoid of oxygen (O_2), however, and in consequence there was no ozone layer to absorb incoming ultraviolet radiation from the Sun and prevent it from reaching the Earth's surface (as it does today). Ozone (O_3) is decomposed by ultraviolet radiation into oxygen molecules and atoms (O_2 and O, respectively), which then recombine to form ozone. Nowadays, there are frequent media reports of stratospheric 'ozone holes' resulting from decomposition of ozone by artificial gases (especially chlorofluorocarbons, CFCs) produced and released since the 1960s, letting in harmful ultraviolet radiation. However, water is a very good shield: it requires a layer of water just a few tens of centimetres thick at most, to absorb all the ultraviolet wavelengths in sunlight.

- How does this help to explain why there was little chance of early life appearing on land in the Hadean?

- Land areas were probably limited in extent and they would have been bathed in ultraviolet light. Land-based lakes might have afforded temporary shields, but they would have been geologically even more ephemeral then than they are now.

Of course, clouds reflect sunlight, including ultraviolet radiation, but we do not know how extensive any cloud cover would have been (see Section 3.6.2, item 2). Another hazard could have been the solar wind (Section 2.3.1). We have no way of knowing the strength of the Earth's magnetic field during the Hadean, so further speculation on this score is fruitless though again it is possible that water could have provided at least partial protection.

In Section 3.1 we saw that there were certainly bodies of water (possibly oceans) at least by the end of the Hadean, also some land areas to provide sediments. Land formed of granitic material (like modern continental crust) must surely have been of small extent – otherwise we should have found more examples of ancient rocks by now. Apart from this, there would have been only islands formed by basaltic volcanoes, which would themselves not be preserved, but would be subducted and recycled back into the mantle.

Question 3.11

Why can we be confident that basaltic volcanoes would be subducted?

Any land areas that did rise above the sea-surface during the Hadean, whether as volcanoes or (towards the end of the period) as embryonic continental (granitic) crust, would also have been prone to abrupt submergence beneath giant waves (*tsunamis*) caused by earthquakes associated with volcanism and subduction, and especially resulting from large impacts.

There is one present-day phenomenon which we can be fairly sure was represented on the early Earth, even quite early in the Hadean wherever there were bodies of surface water. That is **hydrothermal circulation**, resulting from the circulation of seawater through cracks and fissures in hot basaltic rocks where lavas are being erupted to form new ocean crust or to build volcanoes on the ocean floor. Figure 3.16 summarizes the process, and illustrates a simple convecting system with cold water percolating into the crust, migrating towards the heat source, being heated and expelled at the surface. The water exchanges chemical elements with the rocks through which it passes, so that the dissolved constituents in the hot water which emerges are very different from those in the original cold seawater. Many people have seen pictures such as Figure 3.17, which shows the most spectacular manifestation of hydrothermal circulation in the present-day ocean. It is perhaps worth mentioning that geysers and hot springs and mud pools on land today result from the same process, involving groundwater percolation in volcanic areas.

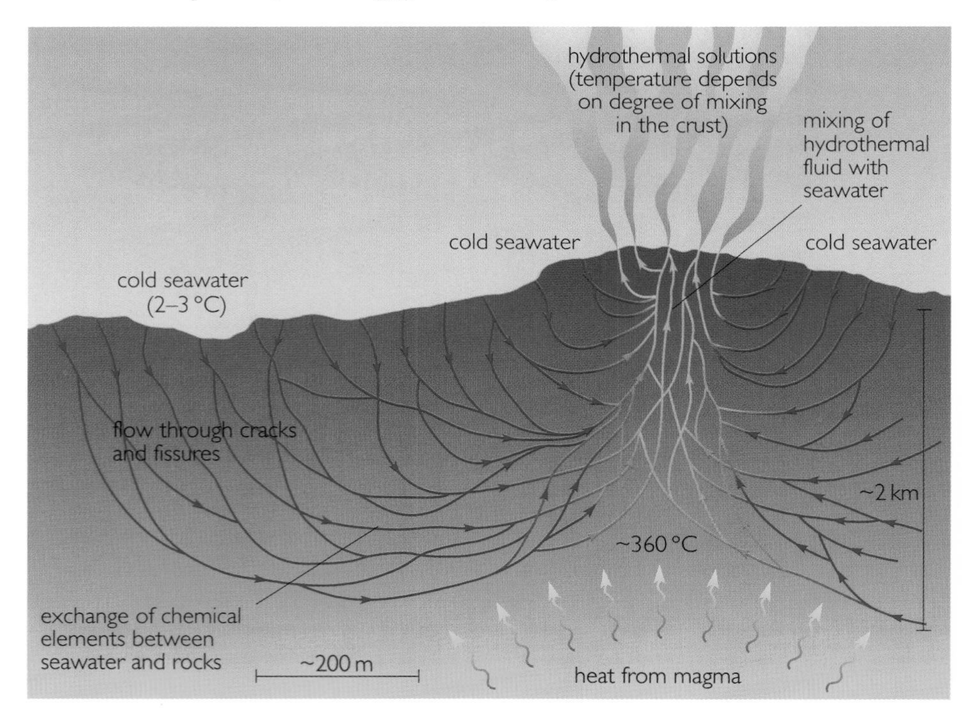

Figure 3.16 Schematic cross-section illustrating present-day hydrothermal circulation at ocean ridges. Flow lines indicate downward and lateral percolation of seawater and its expulsion upwards in a narrow plume, which then mixes with the surrounding seawater.

(a)

(b)

Figure 3.17
(a) A 'black smoker' hydrothermal vent on the sea-bed. The 'smoke' consists mainly of sulfide particles precipitated when the hot vent water (about 350 °C) mixes with cold seawater (about 2 °C). (b) Part of a community of tube worms, one of the commonest animals living round hydrothermal vents in the Pacific Ocean. They are a few centimetres across and commonly more than a metre in length.

- Would we have expected geysers and/or mud pools to be plentiful in the Hadean?
- As any land areas were probably mainly volcanic, they might have been plentiful on those areas. But there was probably not much land on which they could develop.

However, the likely widespread occurrence of hydrothermal circulation in the Hadean ocean has led to suggestions that this process could have been implicated in the origin of life on Earth, a not impossible scenario when you see the variety of life-forms that flourish round modern hydrothermal vents (Figure 3.17), including thermophilic bacteria (Section 3.5).

Throughout the Hadean, the Earth was being continually supplied with organic molecules from carbonaceous chondrites, comets and cosmic dust. As Tables 2.1 and 2.2 show, these molecules would have comprised various combinations of carbon (C), hydrogen (H), nitrogen (N) and oxygen (O) but most would probably have been destroyed in the heat generated by large impacts. However, some would have survived, especially from the rain of interplanetary dust particles; and so the raw materials of life were available at the Earth's surface throughout the Hadean. Can we arrive at a more detailed picture of conditions on that surface?

3.6.2 Hadean climate and weather

It would take too long to recount the evidence, much of it circumstantial, from which scientists have inferred what surface conditions were like on the early Earth, so we shall content ourselves with summarizing the main features.

1 Circulation in atmosphere and ocean

The more or less circular systems of atmospheric anticyclones and cyclones (depressions) you see on weather maps result chiefly from the combined effects of solar radiation and of the Earth's rotation. The former leads to atmospheric convection through heating, the latter gives rise to the **Coriolis effect**. In brief, this causes deflection of the path of anything moving over the surface of a rotating planet which is not frictionally bound to it (such as air and water masses). This deflection is to the right in the Northern Hemisphere, and to the left in the Southern Hemisphere. The distribution of present-day weather systems is thus only partly determined by the distribution of land and sea, and similar systems would develop even on a totally water-covered Earth. Figure 3.18a provides a schematic picture of how such weather systems might look.

Surface current systems in the oceans are driven by the prevailing winds and partly because of this, but chiefly because of the Coriolis effect, they also have an overall *gyral pattern* of circulation, which would have been modified – as in the modern ocean – by the presence of continents, had there been any. Surface currents on a water-covered Earth might look something like Figure 3.18b. It is likely that the Earth was rotating faster in the Hadean (see item 6 below), so the Coriolis effect would have been stronger and the scale of both weather and current systems could well have been smaller.

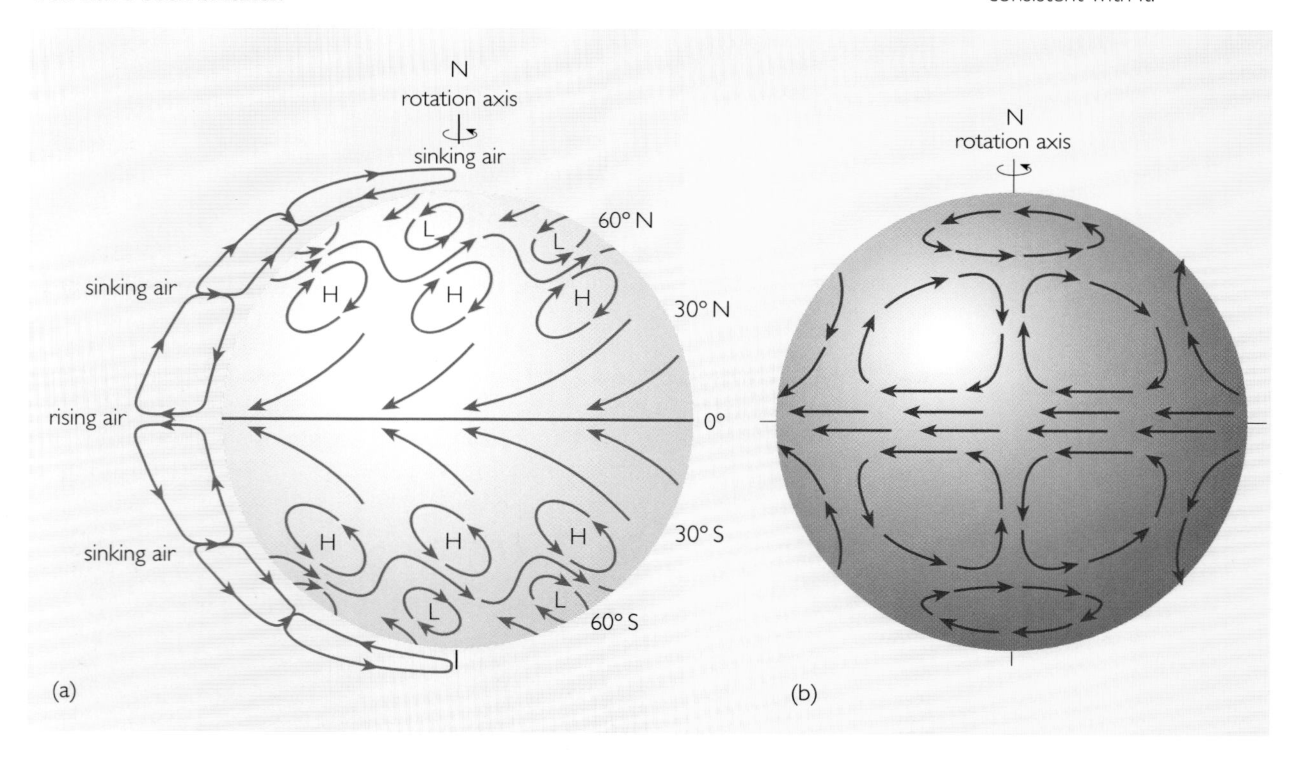

Figure 3.18 Possible pattern of (a) atmospheric circulation and (b) surface current circulation, on a water-covered Earth, with a temperature distribution similar to that found today. H = high pressure system, L = low pressure system. *Note*: Although opposing senses of rotation of circulating systems in both (a) and (b) may seem to contradict the summary of the Coriolis effect given in the text, they are in fact consistent with it.

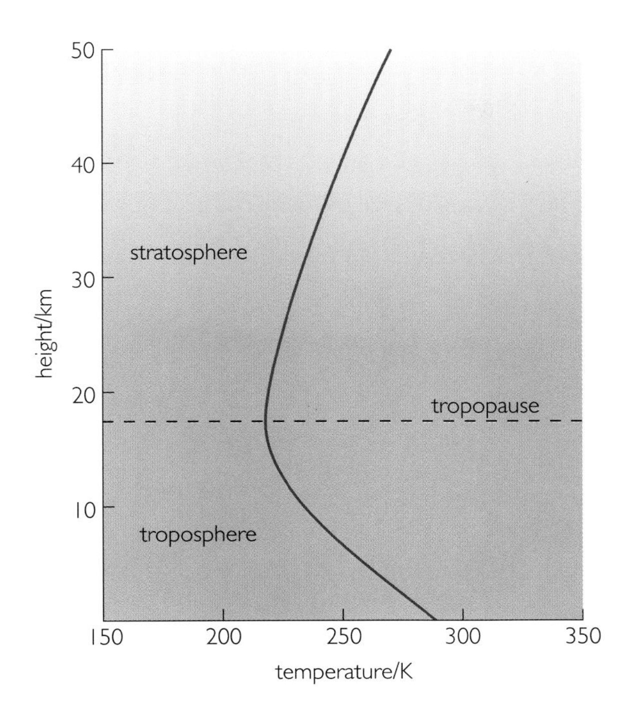

Figure 3.19 Generalized profile to show the distribution of temperature with height in the lowermost 50 km of the present-day atmosphere.

2 Cloud formation

The Sun warms the Earth's surface, which in turn warms the air above it. The warm air rises and cools as it does so – everyone knows that air is colder on mountain tops than on the plains. Water vapour in the rising air condenses to form clouds when the air is cooled sufficiently. The height at which this happens depends upon how much water vapour the air contains, i.e. how 'moist' or humid it is. Condensation releases *latent heat* (Box 2.5) which warms the air and causes it to continue rising. High clouds mark the upper limit of atmospheric convection cells (Figure 3.18a), where the air has lost all its water vapour and become sufficiently cool to be denser than its surroundings and therefore starts to sink back towards the surface. At first sight that might seem puzzling, as we might expect the air to continue cooling with altitude. On the present-day Earth, however, it does not, because of the effect of ozone in the stratosphere above about 20 km altitude. Ozone absorbs ultraviolet radiation and gains energy, some of which is emitted as heat. The atmosphere actually gets warmer and less dense above about 20 km (the tropopause in Figure 3.19), and thus provides an invisible 'lid' for convection. It also prevents water from being carried higher than about 20 km, so the stratosphere is very dry.

- Other things being equal, would you expect the upper limit of convection and cloud formation to have been above or below 20 km in the Hadean?
- Without an ozone layer, the atmospheric temperature could have continued decreasing to much greater altitudes.

Therefore we might infer that cloud formation could have extended to considerably greater heights in the Hadean atmosphere than it does today.

3 Surface temperatures

Was the surface of the Hadean Earth really much hotter than it is now? We have probably given that impression, and it does seem intuitively likely, given continued meteoritic bombardment and rapid convection in the Earth's mantle. However, models of stellar evolution suggest that (as noted in Section 3.1.3) the young Sun (about 4.5 billion years ago) was 25–30% less luminous than it is today, and calculations suggest that a decrease of only a few per cent in the intensity of present-day solar radiation would cause the oceans to freeze over. Even if solar radiation strengthened again, however, it would be difficult to melt that ice, because ice is much more reflective than water, and does not absorb much solar radiation (see Box 3.2). These considerations have led some scientists to conclude that the early Earth was covered with ice that reached thicknesses of 300 m. This is difficult to reconcile with the story in the oldest rocks, which show clear evidence of sediment deposition in liquid water (see also item 4 below). Although we must not forget the existence of micro-organisms living in the dry cold of Antarctica (Section 3.5), it seems intuitively unlikely that life could 'take hold' on an ice-covered planet, Moreover, there are no well-established glacial deposits older than 2450 Ma, and they are of localized extent. There are two ways of resolving this *faint young Sun paradox*. One is that surface heating from impacts and from mantle convection (see item 4) was sufficient to prevent ice formation. The other relates to the composition of the atmosphere.

Box 3.2 Feedback mechanisms

When the Earth formed, the Sun was only about 70–75% as strong as it is now, but the atmosphere was rich in the greenhouse gas, CO_2, which absorbs infrared radiation.

■ What would happen to surface temperatures as solar luminosity increased but the atmospheric CO_2 concentration remained the same?

■ Surface temperatures would rise.

The temperature increase would be accelerated by increased evaporation of water from the oceans, because water vapour is a more potent greenhouse gas than CO_2. The process could continue indefinitely (a 'runaway greenhouse') until there was no water left on the surface – and the atmospheric water vapour would be dissociated by solar radiation (Equation 3.2), most of the hydrogen being lost to space and the oxygen used up in oxidation reactions (Box 1.4). This is an example of **positive feedback**, and it may be the reason why the atmosphere of Venus (which is closer to the Sun than we are, Figure 2.7, and would have started off hotter) is still rich in CO_2 and why its surface temperature is around 460 °C.

The Earth no longer has a CO_2-rich atmosphere and its surface temperature is tolerable for life. What happened to remove the carbon dioxide as the solar luminosity increased?

$$\underbrace{H_2CO_3(aq)}_{\text{carbonic acid}} + \underbrace{CaSiO_3(s)}_{\text{calcium silicate}} \longrightarrow \underbrace{CaCO_3(s)}_{\text{calcium carbonate}} + \underbrace{H_2O(l)}_{\text{water}} + \underbrace{SiO_2(s)}_{\text{silica}} \qquad \text{(Equation 3.3)}$$

Indeed, what happens to keep surface temperatures tolerable? The answer lies in **negative feedback** mechanisms. One of these is the weathering of rocks by the carbon dioxide dissolved in rainwater, which makes it a weak acid, carbonic acid. Expressed in its simplest terms, the reaction is shown in Equation 3.3 where (aq) means aqueous, i.e. in solution, (l) means liquid, and (s) means solid.

The reactions are considerably more complicated than shown above, but the gist of the process is that carbon dioxide is removed from the atmosphere to react with calcium-bearing silicate minerals in rocks, which removes calcium and bicarbonate ions in solution to the sea, where calcium carbonate is precipitated to form limestones. As CO_2 is sequestered (removed), the atmosphere is warmed less, so surface temperatures do not rise as fast. There is less evaporation from the oceans and more condensation to form clouds, which means first that the greenhouse effect of water vapour is less strong, and second that more solar radiation is reflected by the clouds themselves. This is an example of negative feedback, because as temperatures and atmospheric CO_2 concentrations fall, so does the rate of weathering and hence the rate of removal of CO_2 from the atmosphere.

We do not know why this did not happen on Venus – perhaps it did, at first, but the positive feedback was stronger.

■ Might such negative feedback be a solution to the present increase in atmospheric CO_2 caused by burning fossil fuels?

■ Not really, because the time-scale of the increase is decades, and the natural feedback mechanisms operate on geological time-scales of at least millennia and probably longer.

However, external factors may combine to overcome such natural negative feedbacks, and the Earth has been cooled enough to have ice ages. Although the *whole* Earth has not frozen over, the spread of ice from polar regions provides another example of positive feedback. As we saw in item 3 above, snow and ice are highly effective reflectors of sunlight: up to 90% of solar radiation is reflected. Not enough is absorbed to melt the ice, so more ice forms, and the process accelerates. Conversely, however, once the ice has begun to melt, the exposed surfaces can absorb more sunlight, so they warm up and melt more ice, and this process can then accelerate further.

■ Does that have implications for life on Earth at this time?

■ Probably. There *is* global warming, and there appears now to be a consensus that the enhanced greenhouse effect is a contributory factor. But ice formation and melting are seasonal. Each year a little more ice melts anyway, and the positive feedback process outlined above can accelerate the retreat of glaciers and ice-caps and so contribute to further global warming.

It is not as simple as that, of course, because as the world warms there is more water vapour in the atmosphere, which on the one hand may enhance the greenhouse effect further, and on the other may cause more clouds to form, which reflect more sunlight. In the short term we cannot tell which of these trends might prevail, and these are in any case not the only factors to consider. The haze produced by industry and by volcanic eruption also scatters sunlight and can have an overall cooling influence. We cannot with confidence predict the outcome of these effects on human time-scales, but it is important to be aware that natural feedback processes can interact positively or negatively with environmental changes which humans initiate.

4 The atmospheric greenhouse

You have more than likely heard or read about the *greenhouse effect* in the context of *global warming*: human activities are causing rapid increases in the atmospheric concentration of carbon dioxide (CO_2), which is a greenhouse gas (i.e. it absorbs heat, which is infrared radiation). As the Earth's surface is heated by solar radiation, it warms up and emits infrared radiation which is absorbed by greenhouse gases in the atmosphere. Contemporary concerns about human-induced climate change centre round the increasing rates of emission of carbon dioxide from fossil fuel burning and deforestation, which may lead to an enhanced greenhouse effect and excessive global warming. Various lines of evidence suggest that the Hadean atmosphere may have had a few hundred to a thousand times more CO_2 than our present-day one. CO_2 could have constituted as much as a third of the total atmospheric composition at that time – and some scientists believe it may even have been the dominant component. It is at least possible that in the Hadean the 'blanketing' greenhouse effect of all this CO_2 could have combined with the greater amounts of heat coming from inside the Earth (estimated to have been about five times greater than now, Section 3.1.2) to make up for the relatively weak radiation received from the 'faint young Sun'. The lack of an ozone layer would also permit a higher proportion of this radiation to reach the surface.

5 Seawater

The principal dissolved ions in present-day seawater are sodium (Na^+) and chloride (Cl^-). When seawater is evaporated, sodium chloride (common salt, NaCl) constitutes 80% of the resulting salts. The sodium in seawater comes from weathering of rocks and from leaching of oceanic crust by hydrothermal solutions. The chloride originates as hydrogen chloride (HCl) outgassed from the Earth's interior by volcanoes and washed out of the atmosphere by rain. This was no doubt occurring during the Hadean, but with a CO_2-rich atmosphere it is possible that the dominant negative ion in seawater was not chloride but carbonate (CO_3^{2-}) – and the dominant salt formed on evaporation might well have been sodium carbonate, Na_2CO_3.

Hadean oceans could thus have resembled the soda lakes of today (such as those in East Africa) more than the seawater with which we are familiar, and would accordingly have been more alkaline (pH up to 9 or 10, compared with pH 7 to 8 at present). Calcium carbonate (the principal component of limestone) is chemically precipitated from alkaline waters that contain dissolved carbonate ions, especially at elevated temperatures – the calcareous *tufa* deposited from hot spring waters is a good example. This could account for the occurrence of inorganic limestone among the oldest rocks, and it would also be consistent with warm surface conditions in the Hadean.

Concentrations of dissolved carbonate greater than in today's seawater would also be consistent with the occurrence among the oldest rocks of *carbonated* basalts. These contain calcium and magnesium carbonates precipitated by hydrothermal reactions (Section 3.6.1). However, present-day soda lakes also contain dissolved chloride (Cl^-) and sulfate (SO_4^{2-}) ions, just as seawater does. Chloride ions must also have been present in Hadean seawater, because of the outgassing of HCl from volcanoes (see above). There were probably sulfate ions as well, from volcanic outgassing of SO_2 (Section 3.6.1). Calcium and magnesium carbonates as well as calcium sulfate (gypsum, $CaSO_4.2H_2O$) are recorded as having been precipitated from evaporating seawater among 3.5 Ga old sediments (Section 3.1), and it is possible that halite (sodium chloride, NaCl) may also have been precipitated.

6 Days and seasons

The length of our day is determined by the time taken for the Earth to rotate once about its axis. We have used this period to define hours, minutes and seconds. The length of our year is determined by the time taken for the Earth to make one orbit round the Sun (Figure 2.7), so our day and year are thus defined with reference to Earth and Sun – as is the annual cycle of the seasons. Figure 3.20 shows how our winter and summer alternate in the Northern and Southern Hemispheres because of the way the Earth's rotational axis is tilted with respect to the plane of its orbit round the Sun.

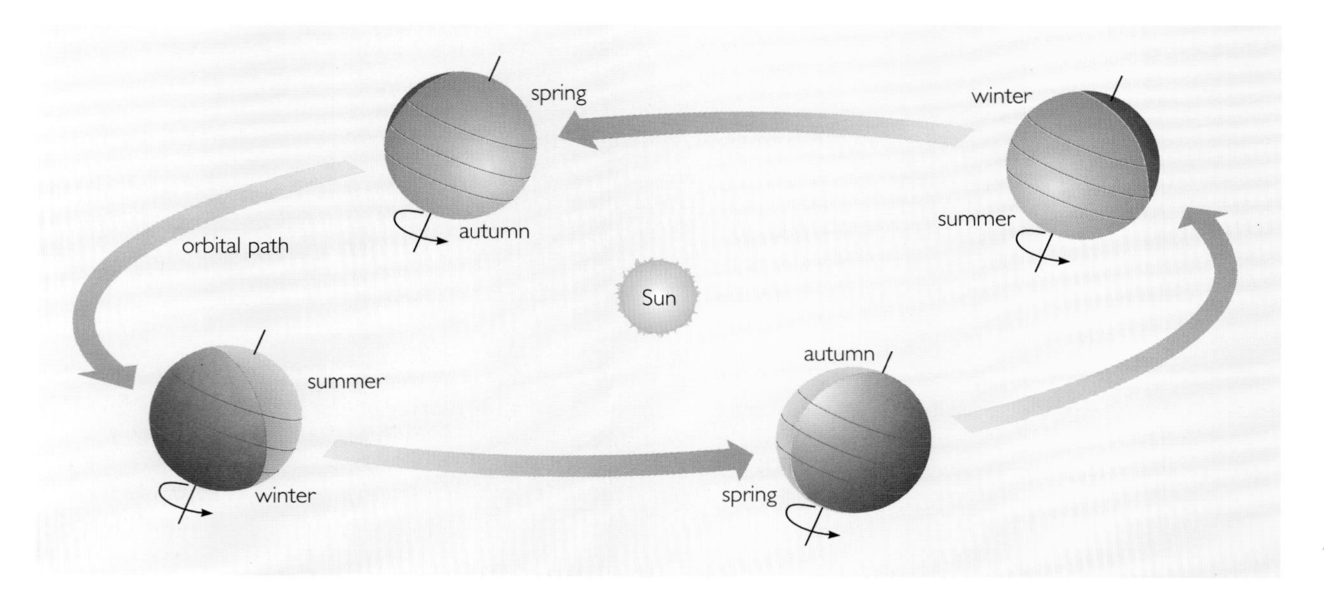

Figure 3.20 Diagram showing how winter and summer alternate in each hemisphere, as the Earth's rotational axis is tilted towards and away from the Sun on opposite sides of its orbit. Spring and autumn occur at the intermediate orbital positions. The present axial tilt is about 23.4° from the 'vertical' (with respect to the orbital plane).

The week and the month, however, are related to lunar motions. The Moon shows the same face to us all the time, because its rotational period (the lunar day) is the same as that of its orbit round the Earth. That orbital period is just over 27 days, nearly four weeks. This is not the place to discuss why the calendar year contains 12 months, not 13, but we should certainly record that the biological rhythms of many organisms (including humans) are finely tuned to lunar cycles; while the cycle of the seasons governs nearly all of life on Earth. It is thus legitimate to ask whether any of these periods was different in the past, and if so, by how much.

- In general, why might we expect all the periods mentioned above to have been shorter in the past?
- The short answer is *tidal friction*, but that needs more explanation.

The rise and fall of tides in the sea result from the gravitational influence of the Sun and Moon, and changes in the relative positions of the Sun and Moon give rise to the fortnightly spring–neap tidal cycle. It is less widely known that tides are raised not only in the ocean, but also in the solid Earth – every day, for example, the interior of the North American continent goes up and down by about 20 cm.

The amount of power exerted by the Sun and Moon in generating the Earth's tides is about 3×10^{12} W, only an order of magnitude less than the rate of energy loss from inside the Earth, 1.7×10^{13} W (Table 3.1). The Moon is the principal 'tide mover' because it is much nearer to the Earth than is the Sun, even though it is much smaller. You may recall the old saying (based on one of Newton's laws of motion) that 'action and reaction are equal and opposite'. So, if the Earth is subjected to tidal stresses by the Moon, the Moon is likewise subjected to tidal stresses by the Earth. The net effect must be that both bodies are acting as remote-control 'brakes' on one another – indeed, the reason why the Moon always presents nearly the same face to us is probably because its rotation period has been slowed to about once per orbital revolution (i.e. once per lunar month). The same must be happening to varying degrees among the other planets (and their satellites), and we may conclude that they must also collectively have some influence on the Sun, despite its huge mass. Without going into details, the net effect of all these changes is to cause the rotation of individual bodies about their axes to slow down and their orbits to expand.

- What long-term effect would these changes have had on day length and the strength of tides on the Earth?
- Days gradually lengthened and tides progressively weakened (tidal ranges became smaller).

Figure 3.21
Fossil of a Devonian coral showing growth bands. Thicker bands separate groups of about 30 finer bands.

Actual evidence of increasing day length is somewhat meagre, and most of it comes from a time long after the Hadean, when life was well established on Earth. Figure 3.21 shows a coral fossil from what is known as the Devonian Period (the geological period from 409 to 363 Ma). It shows growth bands similar to those seen in modern corals, which are known to vary in thickness in response to daily and lunar (tidal) cycles. Microscopic inspection of such corals reveals groups of 30 fine bands on average, interpreted as representing days. Thicker bands separate these groups and are interpreted as representing monthly intervals. The implication is that the Devonian lunar month had 30 days, compared with the 27 we have now.

- If there were more days per month, can we infer that the Earth was rotating faster about its axis in the Devonian than it is today?
- Yes, because the number of days per month is determined by the number of complete rotations the Earth makes about its own axis in the time taken for the Moon to make a complete orbit round the Earth.

The conclusion is that the Earth rotated about its axis 30 times per lunar orbit in the Devonian, not 27 times as it does now. However, the Devonian is almost recent history, compared with the eons of time we have been considering earlier in this chapter. Extrapolations based on astronomical calculations and models suggest that the Hadean day may have been as short as 12 of our hours.

- If the Earth's orbit about the Sun was smaller in the Hadean than now (i.e. it has expanded, see above), what implication might that have had for solar radiation reaching the Earth?
- Solar radiation would have been more intense, thus at least partly compensating for the 'faint young Sun' (Section 3.6.2, item 3). However, we have no means of knowing the magnitude of this effect.

Smaller orbits and shorter orbital periods for the Earth and the Sun on the one hand, and for the Moon and the Earth on the other, could imply shorter years and more lunar months per year in the Hadean. But we cannot confidently quantify these suppositions, nor do we know by how much the angle of tilt of the Earth's axis in the Hadean differed from its present value of about 23.4°. This would affect the distribution of solar radiation at the Earth's surface, and the pattern of atmospheric and oceanic weather systems would not have resembled that suggested in Figure 3.18 (though there would have been gyral circulation because of the Coriolis effect); nor would the pattern of seasonal change have been like that shown in Figure 3.20. We cannot speculate further on these matters here, recording only a consensus that the angle of tilt has changed significantly over geological time, but in a somewhat chaotic manner.

7 A summary picture

Putting all this together, we can conjure up a picture of what the late Hadean Earth might have been like, taking into account the available evidence. We imagine a largely oceanic world, with scattered volcanoes and small areas of continental crust, being weathered and eroded by an atmosphere rich in carbon and sulfur oxides, as well as water vapour, and therefore rather acidic and corrosive. It was warm, if not actually hot, and possibly rather stormy. We can picture large clouds reaching to great heights, causing huge thunderstorms and acting as giant *reflux systems*, cycling water up into the atmosphere by evaporation and returning it as torrential rain to the oceans where its acidity would be neutralized by the alkaline seawater. The ocean itself received both dissolved and particulate products of weathering and erosion, and was already becoming saline, its complement of dissolved constituents being added to and modified by interactions with hot basaltic crust in hydrothermal vent systems.

The rapid rotation rate of the Earth would have intensified the Coriolis effect and made the scale both of atmospheric pressure systems and of gyral oceanic current patterns (Figure 3.18) much smaller than they are today. This could have resulted in winds and currents being stronger than they are now, but that would also depend upon the temperature gradient between the Equator and poles. We know that on the present-day Earth the strength of atmospheric convection (Figure 3.18a), and hence of winds and wind-driven surface currents, is a function of this gradient. If the Earth was more or less uniformly warm in the Hadean, atmospheric convection may have been relatively sluggish, and the winds and currents correspondingly weaker. We have no means of knowing the frequency of Hadean thunderstorms, or whether they occurred against a background of gentle winds blowing over a calm ocean, or one of high winds and storm-tossed choppy seas.

A picture of the climate on a rapidly rotating planet with an atmosphere is in fact available to us. Jupiter is *much* larger than Earth but has a day length of around 10 hours, and must therefore be rotating *much* faster. Figure 3.22 shows that Jupiter has bands of cloud (consisting of hydrogen and methane, *not* ice) parallel to lines of latitude. A closer look at the bands reveals vortices (atmospheric pressure systems) on a variety of scales, which seem to have been active for as long as Jupiter has been observed. Might a picture of the Hadean Earth from space have looked something like this?

While we are on this subject, it is worth digressing briefly to record that because of its enormous size (it is the largest planet in the Solar System, (Figure 2.7) and could accommodate 1300 Earth-sized planets), Jupiter is a kind of cosmic bodyguard for life on Earth: Eugene Shoemaker (of Shoemaker–Levy fame cf. Figure 2.17) has estimated that there would be something like 100 times more impacts on Earth if Jupiter's powerful gravitational field did not deflect comets and asteroids away from us.

Figure 3.22
The storm systems on Jupiter. Jupiter's Great Red Spot is clearly visible.

- The digression should, in fact, remind you that we have missed something out of our depiction of Hell on Earth above. What is it?
- The impact component: Earth continued to be bombarded from space by asteroids, meteorites and comets, and received a continual rain of cosmic dust.

Much of this material contained organic molecules of varying complexity (Tables 2.1 and 2.2). Did these molecules contain the 'seeds of life' themselves, or were they merely the raw materials that were fashioned into complex self-replicating structures in quieter corners of the Hadean Earth by some terrestrial alchemy? In other words, did life *begin* on Earth or was it *imported* from outer space; and was there one 'beginning', or a number of false starts, one 'window of opportunity', or many? The next chapter explores these quasi-theological questions, but from a strictly scientific standpoint.

3.7 Summary of Chapter 3

1. The oldest terrestrial rocks include remains of sediments similar to those forming today and provide evidence that surface geological processes and plate tectonics were operating at the Earth's surface 4×10^9 (4 billion) years ago; though probably on smaller space- and time-scales than today. The sediments contain carbonaceous traces which could be the remains of organisms, so there may have been life on Earth at that time. The resemblance of these oldest rocks to modern analogues makes it plausible to suggest that the Earth's layered structure (iron-rich core, silicate mantle, continental and oceanic crust) had been established 4 billion years ago.
2. Stromatolites (fossilized algal mats formed in shallow marine waters) are known from rocks 3.5 billion years old, showing that life was definitely present on Earth at that time. Organic carbon in the oldest (3.8 Ga old) sediments of western Greenland may represent fossil life-forms. Carbon

isotopes can be used to make inferences about ancient life. Organisms take up lighter carbon-12 in preference to (much rarer) carbon-13, so the $^{13}C/^{12}C$ ratio in fossil organic material is typically lower than in non-organic carbon (i.e. the $\delta^{13}C$ values are more negative).

3 The Earth has a long history of bombardment by asteroids, comets, meteorites, and cosmic dust. Geological processes continually reshape the Earth's surface, so ancient impact scars have long since been obliterated – very few craters older than about 600 Ma are preserved. The Earth itself formed from aggregation and accretion of asteroid fragments, meteorites, comets and interstellar dust, and its likely bulk composition is close to that of carbonaceous chondrites.

4 The average density of the Moon is much less than that of the Earth, so it probably has no iron-rich core. The lunar Highlands are formed of relatively low-density rocks composed principally of plagioclase feldspar and are between 4.5 and 4.4 billion years old. The lower-lying lunar maria are made of denser basaltic rocks and have ages from 3.8 to 3.2 billion years. The favoured theory for the origin of the Moon is that it was torn from the Earth's mantle by glancing collision of another large planetary body some 4.5 billion years ago, i.e. soon after formation of the Solar System, but *also* after the Earth had differentiated into iron-rich core and silicate mantle.

5 The Moon is geologically inert and its surface records a history of meteoritic impacts that spans the Hadean period of Earth's history, the first 600 Ma or so before the oldest rocks were preserved. It provides a 'witness plate' for the bombardment history of the early Earth. The largest lunar maria, some more than 1000 km across, appear to have been formed between 4.0 and 3.8 billion years ago. This is also likely to have been when the Earth experienced its heaviest bombardment.

6 The following inferences can be drawn about conditions on the Hadean Earth, largely from circumstantial evidence.

- Internal convection was more vigorous than at present because of heat energy liberated by decay of a large complement of radioactive elements and by impacts, and from a greater supply of primordial heat (resulting from initial accretion and core formation) than at present.
- The surface was at times sufficiently cool for water to condense and oceans to form (especially at the end of the period, when the sediments in the oldest rocks were formed).
- Hydrothermal circulation would be associated with any volcanic activity in ocean basins.
- Any land areas would have been repeatedly submerged by tsunamis associated with meteorite impacts, if not obliterated by direct hits from the largest bodies.
- Because of the Coriolis effect, weather systems and surface current circulation would have had gyral patterns broadly similar to those observed today, but probably smaller and with greater circulation rates because of the more rapid rotation of the Earth about its axis.
- The atmosphere had no oxygen, so there was no ozone layer to absorb ultraviolet radiation and warm the upper atmosphere. Atmospheric convection and cloud formation may have extended to considerable altitudes.

- Outgassing from the mantle gave an atmosphere probably dominated by nitrogen, carbon dioxide and sulfur dioxide.
- The Sun is thought to have been about 25–30% weaker than it is now, but surface temperatures on Earth may have been kept well above freezing by a combination of heat from the interior and the blanketing effect of a CO_2-rich 'greenhouse' atmosphere.
- The ocean(s) may have been relatively alkaline, like present-day soda lakes and some hot springs, which may explain how original limestone sediments in the oldest rocks could have been (inorganically) precipitated.
- Rotation rates of all Solar System bodies, both axially and orbitally, were probably greater than now (the Earth day may have been as short as 12 hours), but all decreased with time on account of tidal friction arising from gravitational interactions between Sun and planets (and satellites).
- Meteorites, comets and cosmic dust brought organic molecules to the Earth's surface throughout the period. Along with those that survived the initial accretion, they contained the raw materials for the self-replicating molecules essential for life.

Now try the following questions and Activity 3.2 to consolidate your understanding of this chapter.

Figure 3.23
An outcrop of sandstone from Purbeck in Dorset with a herringbone pattern visible in the rock.

Question 3.12
Figure 3.23 shows part of an outcrop of sandstones in Dorset showing a 'herringbone' pattern very similar to that found in modern coastal sediments and which is identified as being caused by the ebb and flow of tides. The sediments shown here are less than 150 Ma old but very similar patterns have been found in sandstones as old as 3.3 Ga. How much support would a herringbone pattern in such ancient rocks give to the hypothesis for the origin of the Moon outlined in Section 3.4?

Question 3.13
Early in the history of the Solar System the Sun was some 25–30% less luminous than it is now, but by way of compensation the amount of energy emanating from the Earth's interior from radioactive decay was about five times greater than now. Use data in Table 3.1 (including note 2) to work out the energy received from the Sun relative to that from inside the Earth at that time, and suggest two other sources of heat that could have lessened the difference.

Question 3.14
It is estimated that the Moon contains only about 25% of the cosmic complement of iron (Fe). Look at Figure 2.11 and assume for the purposes of this question that the abundance of Fe in the Sun represents the 'cosmic complement'. By about how much and in which direction would the point for Fe be shifted if the horizontal scale represented elemental abundances in the Moon?

Activity 3.2

This activity tests both your understanding of aspects of the preceding text, as well as giving you practice at manipulating powers of ten and logarithms, and at plotting and interpreting graphs. If you are not sure of your familiarity with these topics, you should attempt the Activity *and read the answers*.

Figure 3.24 is an interesting diagram. It shows average compositions of the Universe, the Earth's crust and the human body, in terms of relative numbers of atoms, rather than in terms of relative amounts by mass.

We are most accustomed to seeing compositions represented in percentage terms, but there are other ways of doing this. One of them is in Figure 2.11, where the compositions are presented relative to 10^6 atoms of silicon. IMPORTANT: Always remember log scales do not have uniformly spaced subdivisions between successive powers of ten (orders of magnitude).

In the case of Figure 3.24, amounts of elements are expressed as numbers of atoms of each element per 10^5 atoms (i.e. per 100 000 atoms). So if you added up all the numbers in each of the three histograms, the totals would come to 100 000.

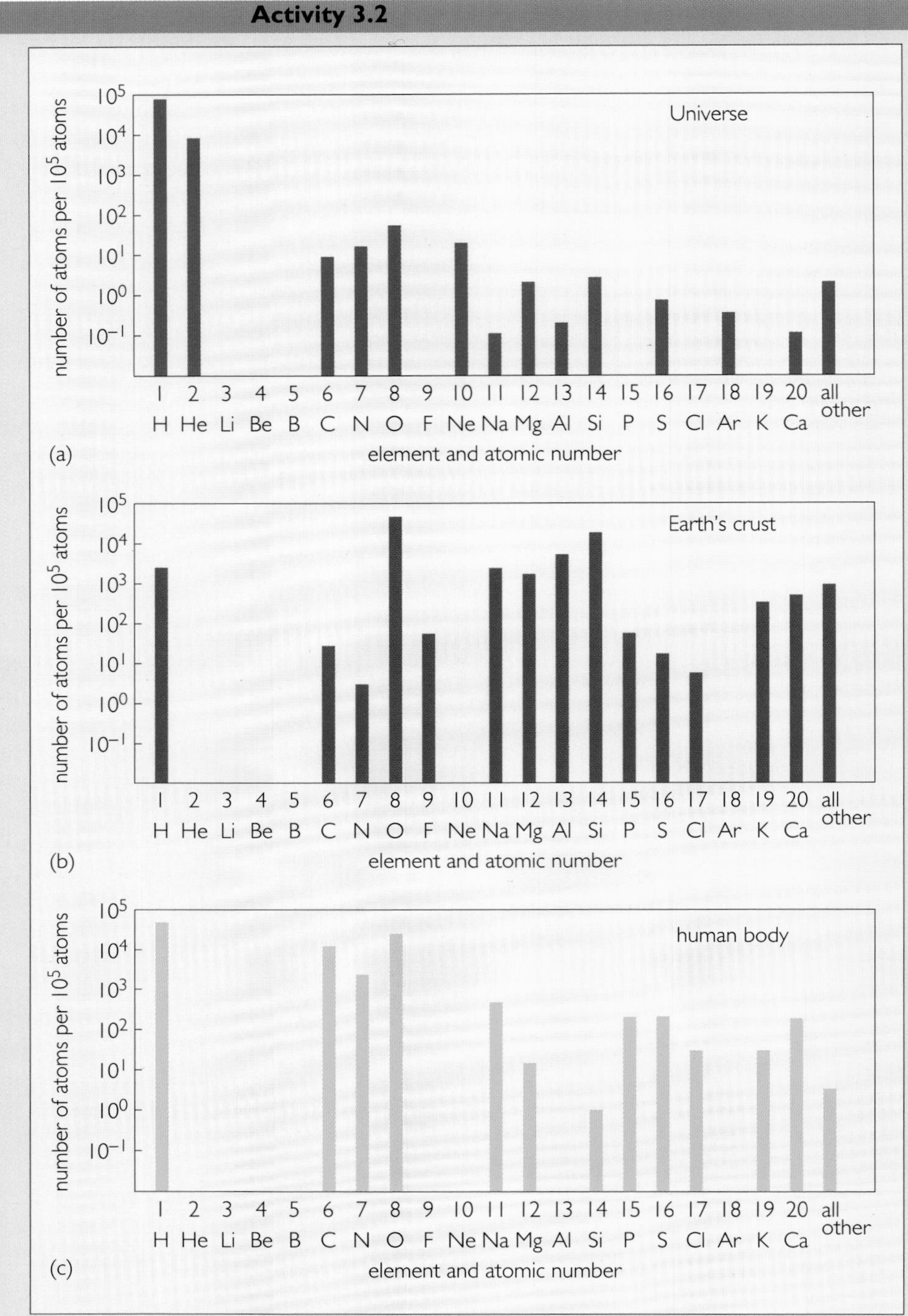

Figure 3.24
Histograms showing average compositions of (a) the Universe, (b) the Earth's crust, and (c) the human body. The heights of the bars represent the number of atoms of each element per 10^5 atoms. (Copyright © 1990 by The Benjamin/Cummings Publishing Company, Inc.)

Questions for Activity 3.2

1 The scales on the histograms in Figure 3.24 are logarithmic, not linear, i.e. each division represents a difference of an order of magnitude (power of ten). Why would it not be realistic to present the data on a linear scale?

2 According to Figure 3.24a, *roughly* how much oxygen is there relative to hydrogen in the Universe (a) in atomic terms, (b) in mass terms? You should give your answers in order of magnitude terms.

3 We have provided you with a logarithmic graph (Figure 3.25) upon which we have plotted most of the data in Figure 3.24. The straight diagonal line represents unity, a slope of 1. You can see the progressive reduction of intervals on the log scales referred to in the preamble to this Activity and the caption to Figure 3.25.

(a) What would you conclude from points that occur on the line in Figure 3.25?

(b) Which two points so far plotted fall closest to that line?

(c) Why are there points along the margin of the graph that have arrows pointing away from the graph?

(d) Which of the elements listed below and plotted on the graph are relatively most abundant in the Universe or the Earth's crust or in the human body? You can use Figure 3.24 to help you answer this question.

hydrogen, H; aluminium, Al; helium, He; silicon, Si.

4 We have plotted all the data for the human body (dots) and data for five elements in the Earth's crust (crosses) onto Figure 3.25. The five elements are hydrogen (H), helium (He), carbon (C), silicon (Si) and aluminium (Al). You should now plot the remainder of the data in Figure 3.24b onto the graph. When you have done that, answer the following questions.

(a) The caption to the original diagram from which Figure 3.24 was taken states that: 'Remarkably, the human body resembles the whole universe in composition more than it does the earth's crust.' Do you think that is a valid statement, from inspection of your completed Figure 3.25?

(b) What is so special about the human body? Would you expect a histogram for the average composition of life-forms on Earth to differ greatly from that in Figure 3.24c?

(c) Does any part of Figure 3.24 bear out the general point that the cosmic processes of elemental synthesis that you read about in Chapter 2 result in the cosmic abundance of the elements being in general inversely proportional to atomic number, i.e. that the heavier elements are in general less abundant than the lighter ones?

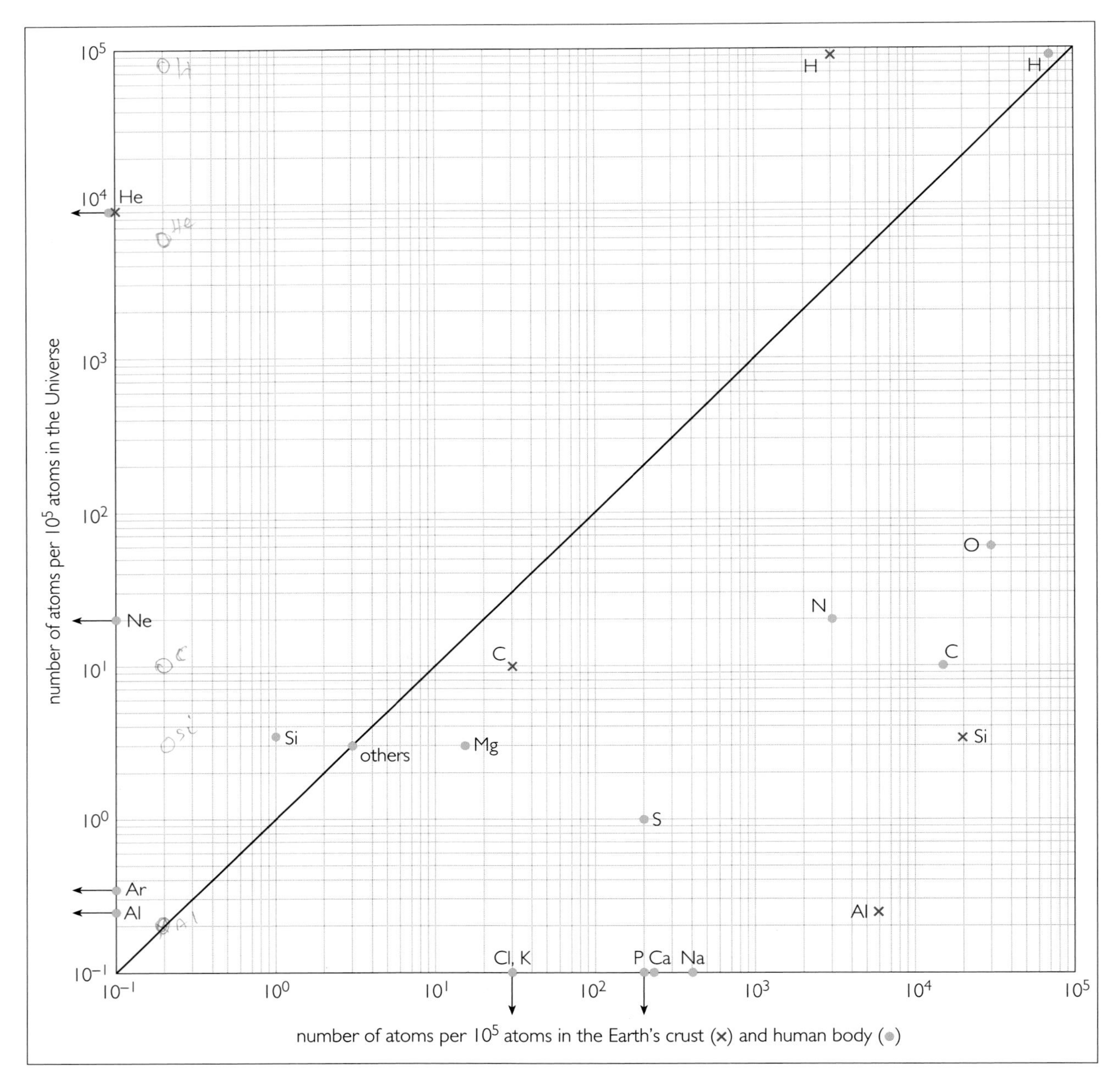

Figure 3.25
Graph for use with Activity 3.2. Note the logarithmic scales: the subdivisions get smaller as you move upwards or to the right within each interval. The straight diagonal line shows where points would fall if either humans or Earth's crust had the same relative elemental composition as the Universe. Data for parts (a) and (c) of Figure 3.24 have *all* been plotted. They are represented by dots, with element symbols alongside. Data for five of the elements in part (b) of Figure 3.24 have been plotted against data in part (a); these are represented by crosses, they are H, He, C, Al and Si.

Chapter 4
The window of opportunity

The cosmic cycle has gone on for billions of years and has churned out the basic chemical elements necessary for life many times. During this huge period of time, stars formed from primordial matter, galaxies evolved, stars died, more stars and planets formed, and life may well have developed on planets elsewhere in the Universe. Why then did a 'window of opportunity' arise on this planet in this Solar System which is a mere 4.6×10^9 years old: about one-third the age of the Universe? In Chapter 3 we saw that the early Earth was subjected to repeated heavy bombardment by material left over from the formation of the Solar System. However, as we discussed in earlier chapters, in spite of the adverse conditions, life was probably present on the Earth around 4 billion years ago – a relatively short time after the Earth itself had formed.

The surface environment of the early Earth was such that it would have extinguished nearly all modern life-forms, although even today there are some specialized bacteria that could have survived, perhaps even thrived in the Hadean. In fact, some features of the early planet would have been familiar to us. For example, solar heating and the Earth's rotation combined to drive atmospheric weather systems (complete with storms, clouds and rain) and oceanic currents. There were volcanoes (and hydrothermal vent systems), even the occasional island, lapped or swamped by waves of various sizes that resulted from winds and tides, as well as from earthquakes and meteorite impacts. Surface temperatures may overall have been rather warm, but may have varied seasonally away from equatorial latitudes (though this would have depended on the amount of axial tilt relative to the Earth's orbit, cf. Figure 3.20).

Question 4.1

There is one crucial aspect of the Hadean environment that would have caused nearly all modern life to be snuffed out like a candle. What 'vital ingredient' was missing?

Much scientific thought on the origin of life is speculative. The very nature of the problem makes it hard to do laboratory experiments to test a hypothesis. So far, scientists have not managed to create life in a test-tube. At various times in the history of the study of the origin of life, scientists have thought they were close to an answer only to be confounded by some new idea or new piece of evidence. By the end of this chapter you will not *know* how life arose on Earth, but you will have a clearer idea of the observations, hypotheses and experiments that have enabled scientists to keep their ideas within the bounds of plausibility.

4.1 Life and the chemical elements revisited

This section builds on the topics and issues raised in Section 1.2, including Box 1.5 on the principal organic molecules of which all living matter is constituted.

- Why could life not possibly have come into being at the time the Universe formed?
- There are several reasons: it was much too hot; only hydrogen was available just after the Big Bang (Section 2.2.3) and elements heavier than hydrogen are essential to life (especially carbon, hydrogen, oxygen and nitrogen, which together constitute about 99% of living material); and there were no planets upon which life could establish a foothold.

This raises a further interesting question: how soon after the Big Bang could life have developed anywhere in the Universe? The answer depends on just how many cycles involving the birth and death of stars would have been necessary to produce enough of the essential elements. Certainly, after the first generation of stars had used up all their hydrogen and helium in fusion reactions, substantial amounts of such elements could have formed and then dispersed in nova and supernova events. Bearing in mind what we know or can infer about the ages of stars, we can surmise that the elements necessary for life might not have become universally available until a few billion years after the Big Bang – though some scientists have suggested that tens to hundreds of millions of years is more likely.

Question 4.2
According to Figure 2.11, where do carbon, hydrogen, oxygen and nitrogen (C, H, O and N) fall in the cosmic abundances of the elements?

The cosmic abundances of the elements generally decrease with increasing atomic number and most of the key elements in biological molecules (Table 1.1) are in the upper half of Figure 2.11. However, complex molecular structures can only be formed from these elements under the right environmental conditions.

Chemical reactions involve interactions between the electrons of atoms, so little chemistry is possible in the heat of stars where elements in a plasma are stripped of their electrons (Section 2.1.1). The atoms must also be close enough to each other to react.

Question 4.3
Use information in Section 2.4.2 to suggest where the density of matter in interstellar space might be high enough for chemical reactions to occur.

The rate of a chemical reaction generally depends upon temperature and in environments as cold as deep space, slow and simple reactions may occur (Table 2.1 and 2.2); but it is difficult to see how molecules as complex as proteins or nucleic acids (Box 1.5) could form. In fact, the only place where life might arise is at the surface of planets. Here we have all the necessary conditions for complex chemistry – a rich mix of elements, in higher concentrations and, at least on some planets, comparatively temperate environments.

4.2 Theories on the origins of life

> *But if (and oh, what a big if) we could conceive in some warm little pond, with all sorts of ammonia and phosphoric salts, light, heat, electricity, etc., present that a protein compound was chemically formed, ready to undergo still more complex changes; at the present day such matter would be instantly devoured or absorbed, which would not have been the case before living creatures were formed.*
>
> *(Charles Darwin, letter to Joseph Hooker, February 1871)*

By the end of the 18th century argument and speculation about how life began had virtually reached an impasse. By that time two main schools of thought had evolved – one that life was created supernaturally, the other that it arose continually from the non-living. Most cultures on Earth contain mythical accounts of a supernatural creation of life. Biblical teaching provides such an account in the opening chapter of Genesis, where we are told that on successive days God brought forth not only the Sun and Moon, but also all living creatures. For centuries following the time of Aristotle, scientists had taken a different view. They thought that one had only to accept simple observations to see that life arose regularly from the non-living: frogs from May dew, algae from clear water, maggots from rotting meat, rats from refuse of various kinds. This view became known as *spontaneous generation.* Scientists like Newton and Descartes accepted spontaneous generation without serious question. Indeed, even theologians (such as the Jesuit, John Needham) could subscribe to this view, because Genesis can be interpreted as telling us not that God created animals and plants directly but rather that He created the correct environmental conditions, namely covering the Earth with waters, to bring them forth.

However, step by step, in a great controversy that spread over two centuries, belief in spontaneous generation was whittled away until nothing remained of it (Box 4.1). As creation theory lay outside science, scientists had no answer to the question of how life arose. For a time they felt some discomfort, then they simply stopped asking.

Box 4.1 The demise of spontaneous generation

Simple observations and experiments, such as those made by the Italian physician Francisco Redi in 1668, disproved many of the original ideas of spontaneous generation. Redi showed that meat placed under a screen, so that flies are unable to lay their eggs on it, never develops maggots. The theory of spontaneous generation lingered on, however, and the debate continued following the discovery of micro-organisms by the Dutch microscopist van Leeuwenhoek in 1683. A century later the Italian Abbé Lazzaro Spallanzani showed that a nutrient-rich liquid (a broth) sealed from the air while boiling never subsequently developed micro-organisms. Needham objected, arguing that the boiling of the liquid had rendered it and the air above it incompatible with life. Spallanzani could defend his liquid: when he broke the seal of the flask to let in new air, it promptly developed microbes. He was unable, however, to show that the air in the sealed flask was compatible with life. Louis Pasteur solved the problem in 1860 with a simple modification of Spallanzani's experiment. Pasteur also used a flask containing a boiling nutrient-rich broth but, instead of sealing it, he drew the neck of the flask out in a long, S-shaped curve. Since its end was open to the air, gas molecules could pass back and forth freely but heavier particles such as dust or moulds in the atmosphere would only rarely reach the liquid. In such a flask the liquid seldom became contaminated, retaining its sterility for long periods. This was just one of Pasteur's many experiments, but he was stubbornly opposed by the French naturalist Félix Pouchet, whose arguments before the French Academy of Sciences drove Pasteur to increasingly rigorous experiments. By the time he finished, Pasteur had shown that the so-called spontaneous generation of micro-organisms was due to the air and fluids not being sterile. He concluded that all living things have their origin in other living things and that there is a total and sharp division between living and non-living. Nothing remained to support belief in spontaneous generation.

4.2.1 Early extraterrestrial theories

Science entered the 20th century without any convincing ideas about how life originated on Earth. Geology and astronomy provided some evidence that the Earth was originally molten and from this it followed that it would have been impossible for life to have existed at the time of its origin. In 1879 Hermann von Helmholtz had proposed that life had not emerged on Earth but had arrived here from another part of the Universe, perhaps on meteorites or comets; the theory was called *panspermia*. As the name implies (germs/seeds everywhere), the theory suggested that life on Earth came from outer space as micro-organisms drifted from planet to planet, propelled by the 'pressure' of star light.

Nowadays, however, the dangers of space travel, which include the effects of radiation, cosmic rays and stellar winds, lethal to nearly all organisms, are generally considered so great as to preclude such large-scale cosmic wanderings. Furthermore, the idea that life is ubiquitous is inconsistent with what we saw in Chapter 2, namely that stars and galaxies originate as clouds of hydrogen or helium. Only as galaxies evolve are the heavier elements necessary for the origin of life formed. For these reasons, panspermia has been discarded in favour of the theory that life originates in those parts of the Universe where conditions are favourable. After all, panspermia does not solve the problem of the origin of life; it merely passes the buck to the vastness of space.

Although extraterrestrial life has now largely been discounted as a source for life on the early Earth, it cannot yet be entirely ruled out. Some bacteria, for example, are incredibly robust, able to withstand extremes of temperature (Section 3.5) and even of radiation. One such bacterium, named *Deinococus radiodurans*, can survive radiation doses thousands of times greater than those survivable by most other organisms, including humans. Equally striking are the bacteria that were trapped with their host insects in amber tens of millions of years ago, which have proved to be not dead, but merely dormant.

An extraterrestrial source of life on Earth still has some eminent proponents, most notably the British scientist Fred Hoyle and his close collaborator Chandra Wickramasinghe. A few decades ago it was the subject of a brief but heated controversy, which started in 1961 with an article in the journal *Nature*. The authors reported the discovery of what they called 'organized elements' in the Orgueil meteorite (Figure 2.10) and claimed that these were microscopic fossil organisms. The far-reaching implications of such a discovery prompted considerable discussion in the pages of *Nature* and at scientific meetings over the following few months. The case against these organized elements being the remains of extraterrestrial life was eventually won by scientists at the University of Chicago, who demonstrated that the organized elements were in fact terrestrial in origin and represented items such as spores and pollen grains that had contaminated the meteorite during its long stay on Earth in the drawers of dusty museums. Box 4.2 has a fuller account of this fascinating tale and of a 19th century hoax aimed at 'proving' panspermia.

Earth is the only place where we *know* that life arose, but the possibility that it evolved independently in other parts of the Universe or even in other parts of our solar system cannot be ruled out (see Section 4.6). As you saw in Section 2.4.2, the Universe is full of many of the building blocks necessary for life. So while life itself may not have arrived on Earth from space, many of the materials required to make it probably did (see Section 4.4.3).

Box 4.2 *The Mystery of the Seeds from Space*

(Slightly edited extract from a talk by Professor W. G. Chaloner, published in *The Listener*, 25 April, 1968, pp. 529–532.)

In an article published in 1961 [in *Nature*], Professors George Claus and Bartholomew Nagy of New York claimed to have observed a number of cell-like bodies from the residues of several carbonaceous chondrites. The majority of these were rather obscure affairs without any obvious distinctive features. But one body from the French Orgueil meteorite was much more intriguing than the others: it is a microscopic regular six-sided object and has three plume-like bulges from alternate faces. Together with a group of other scientists, I examined this object at a meeting in Arizona in the spring of 1962. Claus and Nagy had said of the body in question that it was 'entirely dissimilar to any known terrestrial organism'. In any circumstances this was a rather rash claim to make: many of us who had worked on the microscopic pollen grains and spores of plants were struck by the resemblance of this body to some types of pollen. But a number of workers were excited by these discoveries, and several papers were published illustrating further microscopic bodies dissolved from carbonaceous chondrites. One enthusiastic group even tried to have the international rules that govern the naming of plants altered to allow Latin names to be applied to extra-terrestrial organisms; but this was rejected as being, at best, premature. For the most part these microscopic bodies in meteorites were nondescript items; no two looked alike; few accepted them as direct evidence for the existence of life outside the Earth.

The first serious jolt to this line of thought came when another American worker, Edward Anders in Chicago, showed that the remarkable six-sided structure from the Orgueil meteorite of Claus and Nagy was in fact a pollen grain of ragwort – a weed which, in the United States as in Britain, is a prolific producer of wind-borne pollen. Evidently this pollen had got into their preparations, either when the meteorite was lying in a museum or in the course of working on it in New York. For the same reason it seemed likely that some of the rounded bodies with a nucleus recorded from meteorite preparations might be fungal spores that had got in in the same way. Also it was suggested by another scientist, Mueller, that at least some of the more regularly shaped bodies reported from the meteorites were actually crystals of a greenish mineral, olivine, common to many meteorites.

The most dramatic and bizarre development came just over three years ago [1965]. To pursue his work, the American worker, Anders, had obtained some more of the Orgueil meteorite from France; it actually came from the same museum jar as the fragment on which Claus and Nagy had worked. And on this new Orgueil fragment Anders and his colleagues saw what resembled a small version of the seed capsule of an iris or a lily. There was absolutely no question that, though small, this body (about a fifth of an inch across) was part of the fruit of a plant. It really looked as though these workers had hit upon the fruits of some meteoritic plant. But botanists who examined this fruit quickly realized that the discovery was too good to be true: after comparing the seed capsule with that of some living plants, they were able to show an embarrassingly good match with those of a common species of European rush. But the situation was even worse than this. The rush capsule was well and truly set in the substance of the meteorite – and yet that part of it around the seed capsule proved on closer examination to have been disguised with a mixture of coal dust and glue. It had evidently been carefully attached to the meteorite so as to appear part of it, with a plausible but unmistakably terrestrial cement.

Now who on earth would have perpetrated this strange scientific hoax – this Piltdown from space? Who had so carefully doctored the Orgueil meteorite to equip it with this overly convincing evidence of life? This problem remains almost as obscure as the more genuine aspects of the whole controversy. The question of extra-terrestrial life, like that of the origin of man, is an emotionally charged one. Maybe the same emotional force that led to the Piltdown hoax led someone to supplement the evidence of extra-terrestrial life.

It seems most unlikely that anyone would have perpetrated the trick in recent time: it's much too crude for that. Coal, glue and rush seeds are far too easily identified. Any present-day scientist close to the controversy could have embroidered the evidence from Orgueil in a much more subtle and sophisticated way if he had wished. Anders and his colleagues who revealed the hoax offered a more historical explanation. They pointed out that in April 1864, only a month before the Orgueil meteorite fell, Pasteur read a scientific paper in Paris which conclusively demonstrated that life does not arise spontaneously [Box 4.1]. He showed that a broth, even though exposed to the air, will not develop micro-organisms so long as the spores of such organisms are kept out. Pasteur's claim was that all life, however simple, can only arise from pre-existing life. Now in the eyes of some of his contemporaries, there was an anti-Darwinian element in this claim. Close to the philosophical roots of a Darwinistic concept of evolution is the idea that life itself must have evolved from something that was not life, and that this process was as inevitable as evolution itself: so that life must have arisen many times in the universe, wherever there was a body capable of sustaining it. Perhaps, to some contemporary of Pasteur, the need to demonstrate that life had arisen spontaneously elsewhere than on Earth may have exceeded his respect for scientific integrity. He may have felt that the evidence from the meteorites needed supplementing with some thoroughly unmistakable traces of living matter. We have no idea who this inventive biologist may have been. But unlike the Piltdown hoax, this particular one went undiscovered until it had outlived the context that had brought it into being.

4.2.2 Chemical theories

Chemistry students are often told that when Friedrich Wöhler synthesized the first organic compound in 1828 (Box 1.2), he demonstrated that organic compounds can form without the agency of living organisms. If an extraterrestrial origin for life is ruled out, as well as the supernatural one, just one possibility remains – that life arose on Earth from organic molecules produced **abiogenically** (or **abiotically**), i.e. not manufactured by organisms.

In 1914, a group of scientists suggested that molecules capable of catalysing reactions in the primitive oceans, and possibly also of self-replication, could be formed by spontaneous non-biological processes. The ideas were developed further by the American geneticist Herman Muller, who concluded that the first forms of life on Earth must have been living *genes* capable of mutating and therefore evolving. Muller argued that his 'living genes' gave rise to photosynthetic organisms, the ancestors of modern plants. However, most scientists now agree that no single molecule or gene can ever be 'alive' outside of a cell; so it is essential also to explain the emergence of membranes which provide the walls of cells.

In November 1923, a young Russian biochemist, Alexander Ivanovich Oparin, published a book with a different hypothesis. He had observed that when an oily liquid is mixed with water, the two sometimes form a stable mixture called a *coacervate* (a kind of emulsion), with the oily liquid dispersed into small droplets suspended in the water (Figure 4.1). Coacervates are easily formed by non-biological processes and the droplets have a certain superficial resemblance to living cells. Oparin suggested that life began by the successive accumulation of more and more complicated molecules within the droplets. The physical framework of the cell came first, provided by the naturally occurring droplets. Enzymes (natural catalysts, Box 1.5) came second, organizing the random population of molecules within the droplets into self-sustaining metabolic cycles. Genes came third, presumably because Oparin had no idea what they were and only a vague idea of their function, but he believed they appeared to belong to a higher level of biological organization than enzymes.

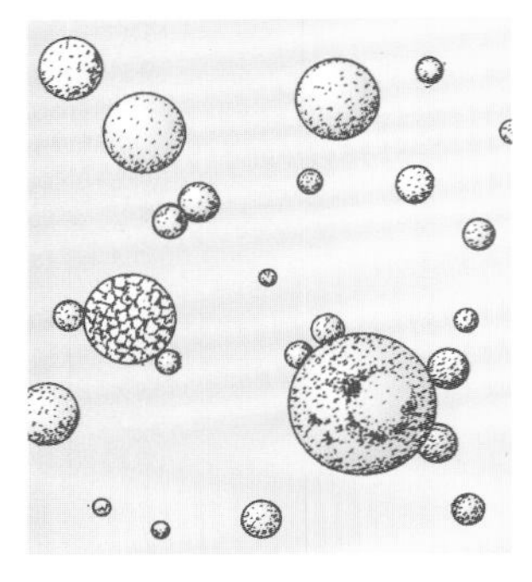

Figure 4.1
Highly magnified sketch of coacervate droplets similar to those formed when an oily liquid is mixed with water. The smallest droplets are of the order of a few nanometres (10^{-9} m) in diameter.

Contemporaries of Oparin thought along similar lines. The British biologist J. B. S. Haldane argued, in 1929, that the origin of life could be explained in a context of *chemical evolution*. The main difference between Oparin's and Haldane's views was that whereas for Oparin the first living things were cells, for Haldane the first living things were viruses, the simplest of all organisms. However, this idea is fundamentally flawed, because viruses are not capable of reproducing themselves unless they are inside a host cell and Haldane's original ideas would not explain the origin of the cell.

You need to bear in mind that the Oparin–Haldane theory was suggested at a time when little was known about cell metabolism and nothing at all about the nature of genetic material. This model of chemical evolution was generally accepted by scientists for 40 years largely because, although there was no evidence to support it, there were no scientific alternatives. Everything had to be guesswork until biochemistry and molecular biology had progressed far enough to enable scientists to handle problems such as the chemical composition of proteins and nucleic acids and the manner in which genetic information is stored. Eventually, scientists began to test the Oparin–Haldane thesis (including the coacervate hypothesis) and in recent decades there have been great advances in the understanding of these matters, as you will see shortly. But first, we need to explain some basic genetics to make the rest of the chapter easier to understand.

4.3 Life's information superhighway

In Chapter 1 we examined how we might decide whether or not something was 'alive', and the ability to self-replicate was clearly one essential criterion. In this section we will consider the task that faces a simple single-celled prokaryotic organism (see Box 4.3). There is good evidence to show that the first organisms to evolve on Earth were prokaryotes. For any organism to reproduce and to continue doing so, it has to replicate itself countless times with minimal error; which is a truly staggering feat even for the simplest prokaryote.

Box 4.3 Prokaryotes and eukaryotes

With a few exceptions, e.g viruses, all organisms are built of cells. One major group of organisms has cells with the genetic material contained within a membrane to form a well-defined **nucleus**. These are called **eukaryotes**. Eukaryotes may be complex multicellular organisms, e.g. plants and animals (including us) or may consist of only simple organisms such as dinoflagellates, foraminiferans (known as protistans) or unicellular algae. Organisms whose cells lack a membrane surrounding the genetic material, i.e. lack a nucleus, are called **prokaryotes**; a name virtually synonymous with bacteria. Bacteria thrive in all manner of strange environments (including human bowels and the extreme conditions outlined in Section 3.5) and were the earliest forms of life on Earth. All prokaryotes are single-celled (though some can form colonies) and they can be subdivided on the basis of their DNA (there are few visible differences) into two Kingdoms – the Eubacteria and Archebacteria (or Archea), the latter being the most ancient life-forms of all.

So how *does* a simple single-celled organism achieve self-replication? In Box 1.5 you encountered a group of complex molecules called the *nucleic acids*, remarkable molecules that are even larger than *protein* molecules, with relative molecular masses that run into the billions. Like proteins they are polymers, made up of simpler units, and in the case of the nucleic acids these simpler units are called *nucleotides*. Each nucleotide consists of three subunits: a sugar, a phosphate group and a nitrogen-containing base (where 'base' is used in the sense of being opposite to acid, rather than in the sense of a foundation). The sugar and phosphate groups act as alternate links in a chain joining the nitrogen-base units together. The nitrogen bases themselves stick out from the side of the chain, with one base attached to each sugar (Figure 4.2).

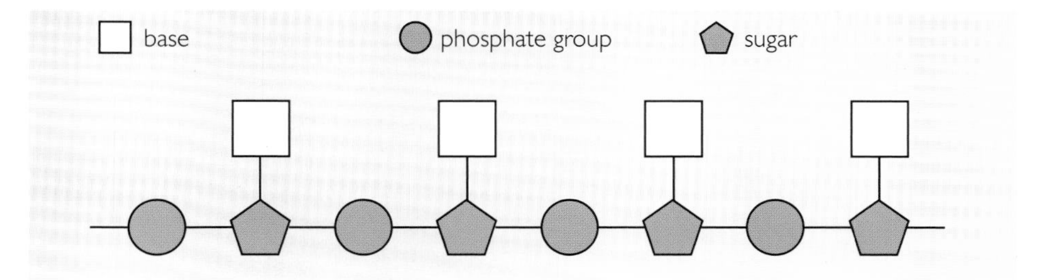

Figure 4.2 Schematic structure of a nucleotide illustrating the bases attached to the chain of alternating sugars and phosphate groups.

There are two distinct kinds of nucleic acid, each of which has a critical role to play in life. One is called ***d**eoxyribo**n**ucleic **a**cid* or **DNA**, after the sugar unit it contains, called *deoxyribose*, which is made up of five carbon atoms. The other is ***r**ibo**n**ucleic **a**cid* or **RNA**, after the sugar unit called *ribose*, also containing five carbon atoms. Attached to each sugar in the nucleotides making up DNA is one of four possible nitrogen-containing bases, *cytosine, thymine, adenine* and *guanine* – which are often abbreviated to the single letter codes C, T, A and G. The first two of these bases (C and T) belong to a class of compounds called pyrimidines, the second two (A and G) to a class called purines. In RNA, thymine is replaced by a different pyrimidine, *uracil* which is given the code U. DNA and RNA differ in one other important respect: whereas RNA consists of a single chain of alternating sugars and phosphate groups (and bases), DNA is in fact made up of *two* such chains, which are twisted around each other in a structure resembling a spiral staircase – the famous **double helix**. At each sugar, a purine or pyrimidine base juts out from the chain into the 'stairwell' where there is a weak attraction (hydrogen bonding) between the bases, thereby holding the two chains together (Figure 4.3).

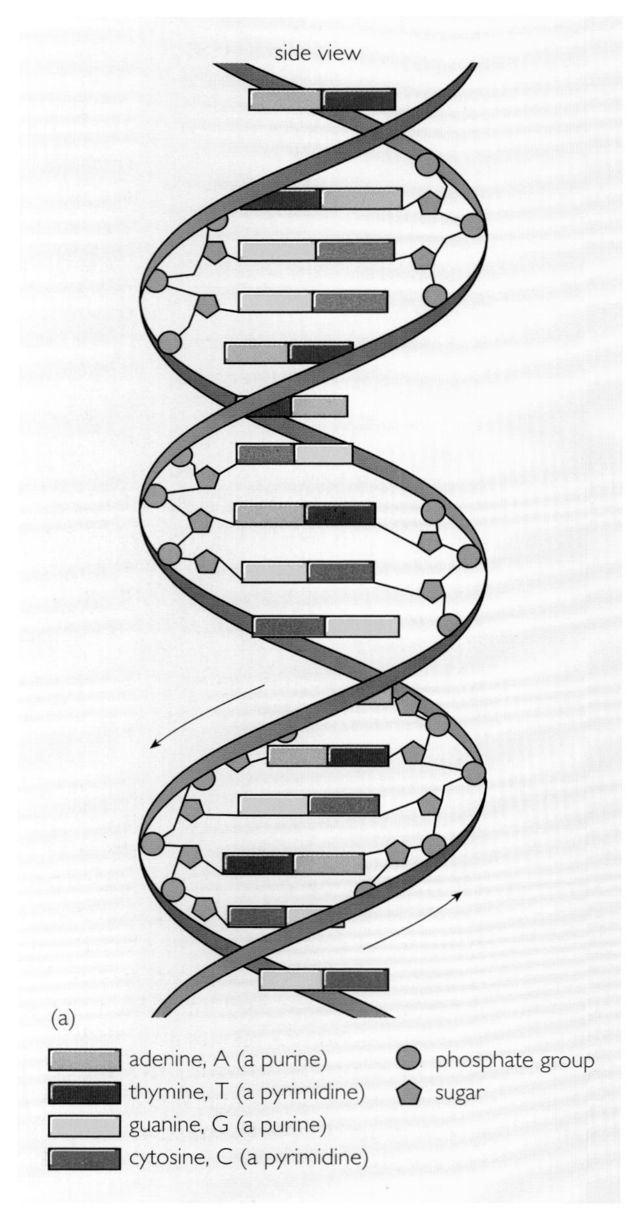

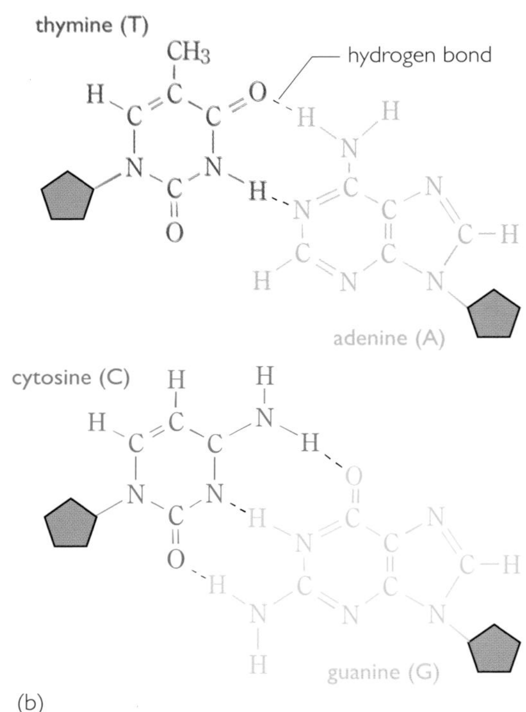

Figure 4.3
(a) The double-helix structure of DNA (deoxyribonucleic acid), which consists of two polymers linked together by pairs of purine and pyrimidine bases according to the A–T, G–C pairing rules: adenine pairs with thymine (A–T), guanine with cytosine (G–C). The twisted green 'ribbon' illustrates the chains of phosphate groups and sugars which make up the backbone of the double helix. (b) Two base pairs are shown illustrating the hydrogen bonding. There are two hydrogen bonds between the A–T pair and three between the G–C pair.

The double helix structure of DNA was first elucidated by Francis Crick and James Watson in 1953 and they were subsequently awarded a Nobel Prize for their work. A key aspect of this structure is that on the one hand there is insufficient room across the 'stairwell' for two purines to occupy the same level, while on the other, there is too much room for two pyrimidines (which are smaller molecules than the purines) to meet and join together. Moreover, the molecular structure of the bases is such that adenine pairs with thymine (A–T) and guanine pairs with cytosine (G–C). Thus the two strands of the DNA molecule are complementary to each other, so if we know the sequence of bases in one strand, we know the sequence in the other.

To Crick and Watson the double helix structure and the complementary nature of the two strands suggested a possible copying mechanism, a means of self-replication. Other scientists soon realized that within these molecules resides not only the information that controls the structure and activities of all organisms, but also the machinery for passing on this information, the **genetic code** (see Box 4.4) – the 'blueprint for life'. But how is the information passed on?

Box 4.4 The genetic code

(This is a condensed summary and you are *not* expected to understand the details.)

Cells use DNA to make proteins but the production of proteins does not proceed directly from DNA. Information in the DNA sequences of the **genes** is translated into amino acid sequences of proteins using three forms of RNA as intermediates. The process is called *translation* and uses transfer RNA (tRNA) to convey individual amino acids to the messenger RNA (mRNA) template, where they are 'clipped' together by ribosomal RNA (rRNA) to form proteins. The nitrogen-containing bases of the DNA molecule are read in groups of three, and it is this trio of bases that, after translation via mRNA, specifies which of 20 possible amino acids are placed at a specific position in the protein. There are even 'start' and 'stop' codes that determine the size of the protein and specific pairs of bases that are copied in the initial *transcription* of the code to mRNA. The pairing 'rules' are the same as for DNA except that the base thymine is replaced with the base uracil. Thus, the mRNA code GGA, for example, specifies the amino acid glycine (Figure 1.5c see also Figure 4.10), while the code UUU specifies the amino acid phenylalanine (see Figure 4.10). Given that there are four DNA bases grouped in trios, there are 64 ($4 \times 4 \times 4$, i.e. 4^3) possible codes. This is more than sufficient to specify each of the 20 possible amino acids and the 'start' and 'stop' codes, while some amino acids are specified by more than one code.

If you were to make a photocopy of this page, then a photocopy of the photocopy, and so on, after about 10–15 copies the page would become blurred and almost unreadable. The information is being transcribed from one generation to the next in an *analogue* fashion with many variables affecting the quality of the copy, such as the quality of the lens on the copier, the amount of toner, the amount of heat used to fuse the toner to the paper, and so on. However, the original text was stored *digitally* on a word processor in binary code, that is, a series of ones and zeroes. A *digital* mechanism for copying is much more reliable since a very limited amount of information has to be transcribed – a one or a zero – and this results in a much higher quality copy. It may come as something of a surprise to realize that life on Earth perfected a digital style of copying nearly 4 billion years ago. It used the A–T G–C code of DNA to pass information from one generation to the next, as well as to control the processes taking place within an organism.

Here is an example. The sequence of bases in a typical segment of one strand of a DNA molecule might be as shown in Figure 4.4. We can write the code for this segment using the notation outlined above, as

—A—C—G—G—T—G—G—

- What will be the code for the complementary strand to this segment in the DNA molecule and why?
- It must be: —T—G—C—C—A—C—C—

  This is because adenine (A) can only pair with (complement) thymine (T), and guanine (G) always pairs with (complements) cytosine (C).

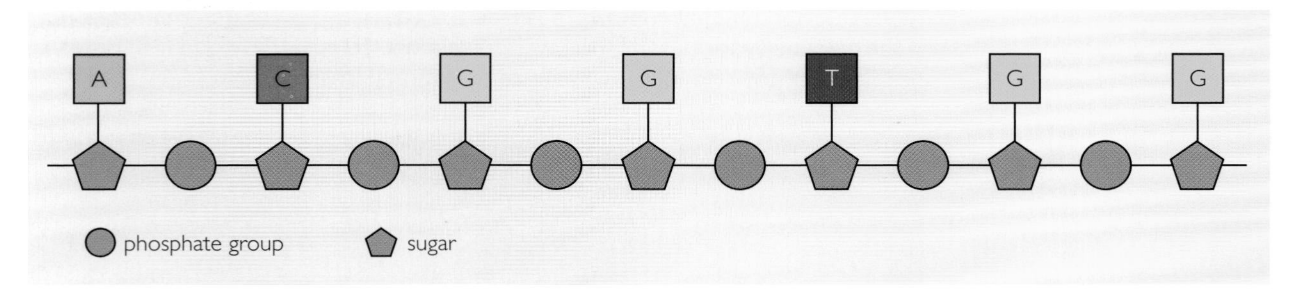

Figure 4.4 Schematic diagram showing the sequence of bases in a typical segment of a strand of DNA. The chemical structures of the four bases A, T, G and C are shown in Figure 4.3.

Such a mechanism obviously has several advantages for a living organism, not the least of which are simplicity and accuracy. DNA replication enables information to be faithfully copied from one generation to the next. During cell division the DNA double helix (Figure 4.3) 'unzips', splitting the molecule along its length into two separate strands (Figure 4.5). Each of these strands acts as a template to which new nucleotides are attached, so forming two new double helices which are identical to the original.

- Has the copying of the genetic code across the generations through geological time been flawless?
- No. Life on Earth has become enormously diverse throughout its history (especially in the last 600 million years or so). This could not have happened had the copying of the A–T G–C code been perfect.

Errors do occur, albeit at very low frequencies and these are called *mutations*. In the longer term, mutations are the source of the variability upon which natural selection has acted, giving rise to the great biodiversity we see upon Earth today.

The genetic code described in Box 4.4 is virtually ubiquitous (the rare exceptions are a special kind of DNA called mitochondrial DNA and that found in certain viruses): practically all life on Earth uses the same nucleotide codes to assemble proteins from the various amino acids. Furthermore, since the strands of DNA are of indefinite length, the range of variation is effectively infinite. This is the source of the genetic diversity of life, and its appearance on Earth nearly 4 billion years ago represented an information explosion. DNA, it seems, is not just the thread of life but the original 'information superhighway'.

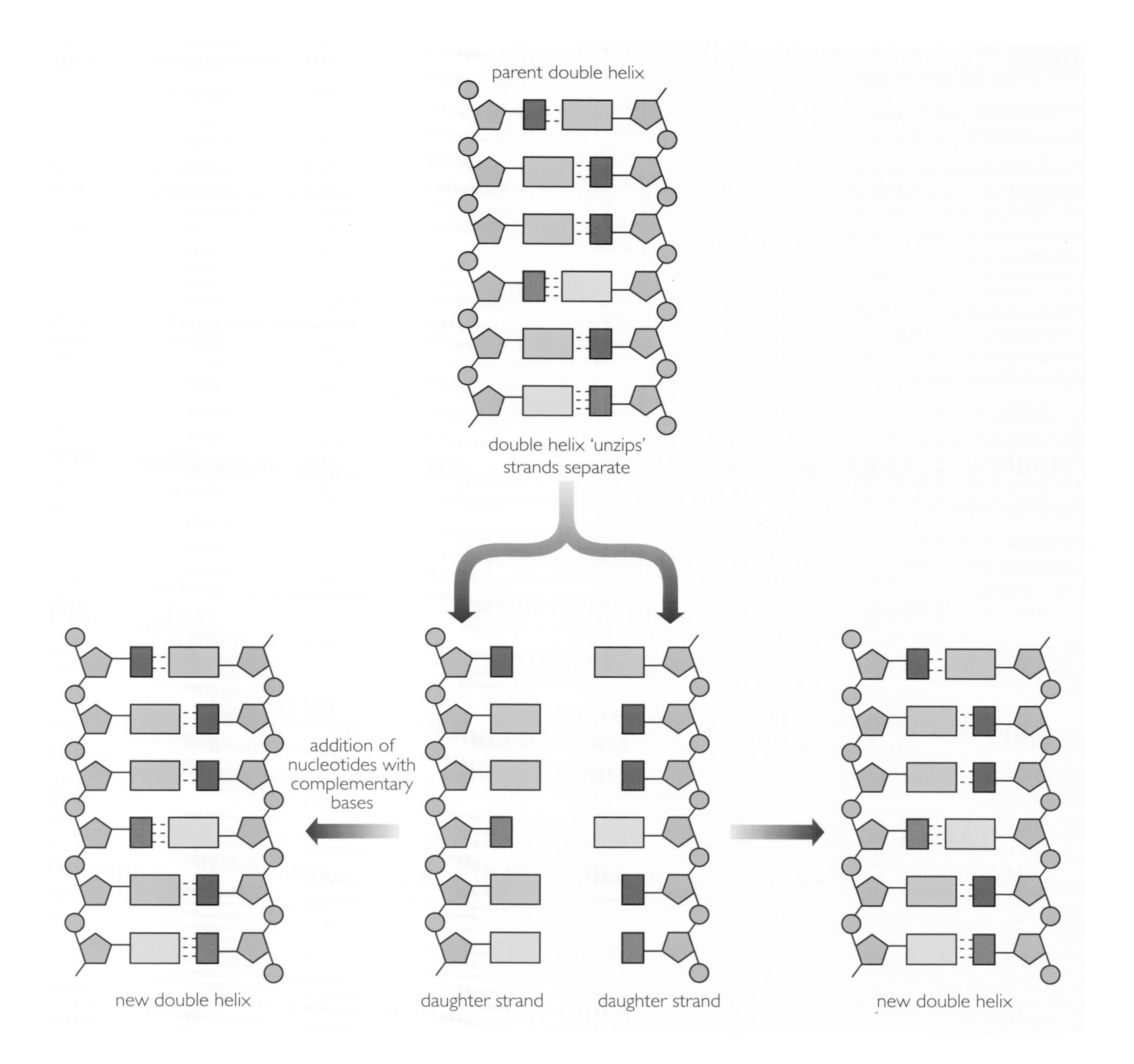

Figure 4.5
Schematic diagram representing DNA replication. Two daughter strands are formed as the double helix 'unzips'. These strands act as templates to form two new double helices. The boxes used to represent A and G are larger than those used for C and T because purine molecules are larger than pyrimidine molecules.

- Aside from some minor exceptions then, all the available evidence indicates that the genetic code is common to all life on Earth. What does this imply about the origin of all life on Earth?

- It implies that all of the species that have existed on Earth for the last 4 billion years or so have a common ancestor. The genetic code is the molecular-biological key to evolution.

So far we have been concerned with the situation for the simplest single-celled organisms, the prokaryotes. Perhaps the second most important event in the history of life on Earth – after its origin – was the appearance of more complex organisms, the eukaryotes (Box 4.3). The eukaryotes raise some interesting questions about how DNA operates in cells. Cells in a multicelled eukaryote are not all the same, yet all such cells contain the same genetic code. Why then are the cells in your skin different from those in your bones? Furthermore, how do the cells that form your nose know that they should produce the shape of a nose and not that of a finger? The answer lies in the ability of part of the genetic sequence in the DNA of a particular cell type to switch on or off at just the right moment.

Life's information superhighway, in which the data in DNA are transcribed into RNA, and RNA makes proteins (Box 4.4), is in essence based on the five nucleotide bases (codes A, T, G, C and U), which seem to have existed as long as life on Earth. Despite all the evolutionary changes and diversification that have taken place they have remained the same and are found in very nearly all forms of life on Earth (Figure 4.6). But which came first – DNA or RNA? We will revisit this question in Section 4.5.1, after we have looked at how early ideas about the origin of life have fared in the light of scientists' improved understanding of genetics and of the likely conditions on the early Earth.

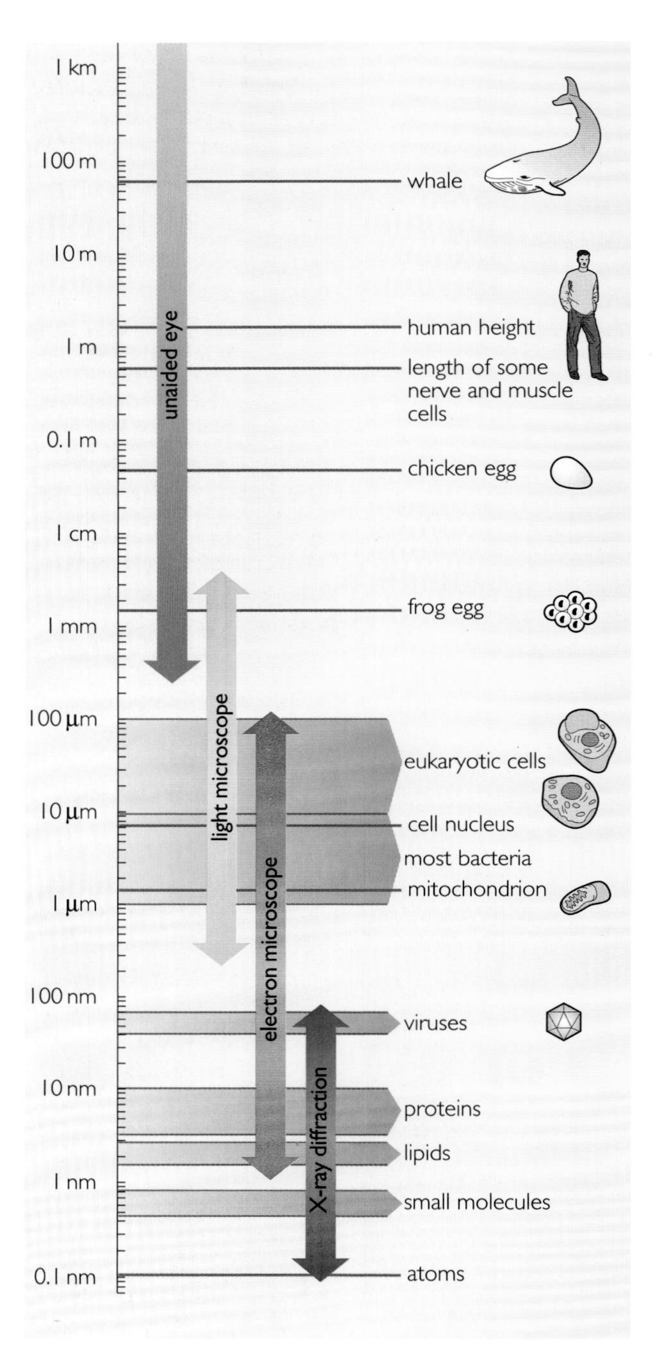

Figure 4.6
The complexity and diversity of living organisms becomes apparent when we see the range in size from the basic molecular building blocks to the largest organisms. The scale is logarithmic. (Copyright © 1990 by The Benjamin/Cummings Publishing Company, Inc.)

4.4 Building blocks of life

The complexities of life processes require many of the molecules that participate in them to be of enormous size, especially the proteins (e.g. Figure 1.6) which require vast numbers of amino acid building blocks. For example, the DNA molecules released from *human* chromosomes have a long threadlike shape and would be some *4 cm in length* if stretched out (Figure 4.7). This thread corresponds to just two molecules, each containing billions of amino acids. These giant molecules are the 'magnetic tapes' from which the genetic code is read. Since the amount of information needed to specify the structure of a multicellular organism is vast, not only are these 'tapes' extremely long but they must also be extremely accurate. In Section 4.3 we used the analogy of a photocopy to highlight the problems of the analogue method of copying information. Life has to overcome the same problems and as we have seen, DNA is a molecule that takes the digital age back billions of years. It can copy itself exactly, indeed it *must* copy itself exactly most of the time, or life would rapidly get into problems. Yet as we saw in Section 4.3, errors do occur and every now and then an inaccuracy is transcribed. Some errors arise by chance, others may be caused by external changes in the organism's environment. Whatever their cause, it is these mutations in the genetic code that are responsible for biodiversity and for life's ability to evolve and to survive sometimes drastic changes in the Earth's environment.

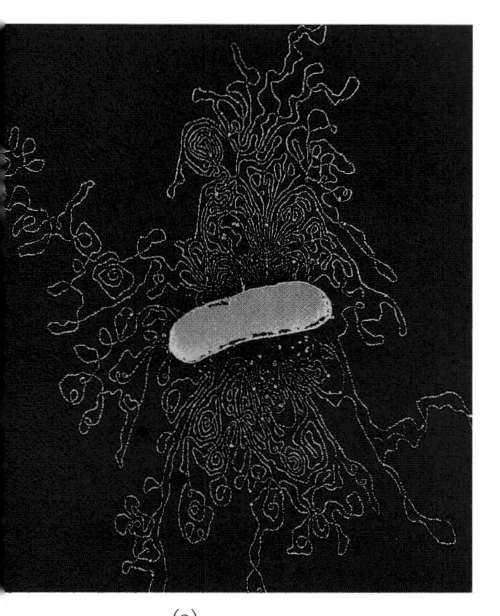

(a)

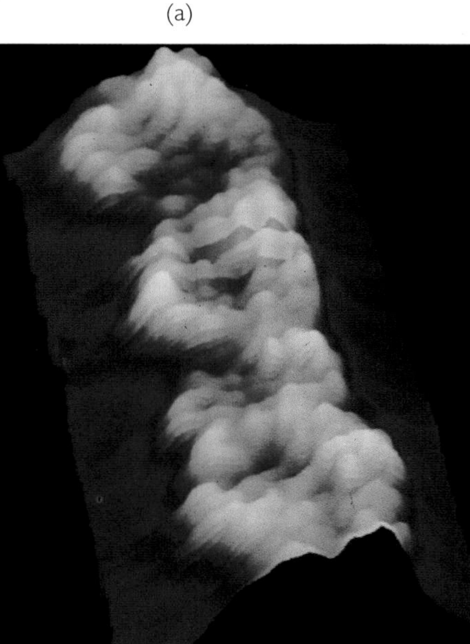

(b)

Figure 4.7
(a) False-colour transmission electron micrograph of the bacterium *E. coli* surrounded by its DNA. The *E. coli* was treated so that its DNA was ejected from the cell. The DNA has a total length of 1.5 mm, i.e. it is 1000 times the length of the cell. The DNA released from *human* chromosomes would be about 4 cm in length.
(b) A scanning electron micrograph of a segment of a DNA molecule magnified many millions of times.

The synthesis of such large molecules poses a challenge to the cell of any organism. If the cell were to behave like an organic chemist in the laboratory carrying out a complex synthesis bit by bit, millions of different types of reactions would be involved. Instead, cells adopt a more efficient modular approach for constructing these molecular giants, much in the same fashion as building a prefabricated house. The 'modules' are polymers, made by joining together a fairly limited range of prefabricated units linked together by identical mechanisms. However, here we have an interesting dichotomy. We know that a modern cell can synthesize large and complex molecules. To do so it uses the pre-encoded instructions in the DNA molecules, many thousands of times over. Yet, how do we make a cell without first having the instructions in DNA? Clearly, the code for life had to come first, because without it there could be no life. In what form this code first appeared, and how it was put together from the basic building blocks, are critical to our understanding of how life may have originated on Earth.

In the next section we will examine how Oparin's and Haldane's ideas of chemical evolution (Section 4.2.2) have been modified and tested over the last 40 years and where and how the basic building blocks of life may have formed.

4.4.1 'Warm little ponds'

The main approach to unravelling life's origins adopted by scientists in the second half of the 20th century has been to see what chemical reactions occur when conditions are set up (in the laboratory) that are thought to be similar to those that existed on the primitive Earth. Inferences about the likely composition of the atmosphere of the early Earth are obviously important (Section 3.6.1), but another factor that must be considered is the possible energy sources available for **pre-biotic** (or **pre-biogenic**) chemistry on the early Earth (i.e. the chemistry that existed before there were life-forms). As you established in Question 3.13, by far the largest energy source, then as now, was the Sun, far exceeding the energy from the Earth's interior, even though the Sun is thought to have been some 25–30% weaker than it is today. However, it is the type of radiation that is important, not the amount of energy, because only radiation with a wavelength of less than 200 nm can photolytically dissociate simple compounds like water and methane (H_2O and CH_4, cf. Section 2.1) into reactive components and initiate the synthesis of larger organic molecules. It is estimated that only around 1% of ultraviolet radiation was of this type, but it was still probably the major source of energy available for synthesis of organic molecules on the early Earth.

Question 4.4

(a) In Question 3.13 we estimated that the total solar radiation reaching the Earth's surface in the Hadean was of the order of 10^{17} W. It is known that ultraviolet radiation makes up only about 20% of the total solar radiation, so what was the solar power available for pre-biotic chemistry (i.e. from radiation with a wavelength of less than 200 nm)?

(b) The Sun is much stronger now than it was in the Hadean, but little ultraviolet radiation reaches the Earth's surface today. Why is this?

It is possible that short pulses of intense energy, e.g. from electric discharges (lightning) could be more important for abiotic synthesis than a steady flux of solar radiation. Lightning discharges may have been more frequent in the Hadean atmosphere than they are today, if our inferences about giant cloud columns are valid (Section 3.6.2, item 2), and their total contribution may have been of a similar order of magnitude (i.e. about 10^{14} W). Smaller sources of energy included heat from the Earth's interior (through radioactive decay of unstable elements and isotopes such as uranium and potassium-40), but we know that this was still three orders of magnitude less than that from solar radiation (Question 3.13) and would in any case have been mainly in the longer wavelength infrared and so of relatively minor importance for pre-biotic chemistry. Related to this internal heat flow, however, was abundant volcanic activity, with eruption of lavas at temperatures well over 1000 °C, which is quite hot enough to dissociate methane, for example. The vast amounts of heat liberated by meteorite impacts could have had a similar effect, but as this energy was produced virtually instantaneously, it would be more likely to destroy any life-forms than to encourage the construction of complex molecules.

Early experiments

The scenario for chemical evolution proposed by Oparin in his 1924 book *The Origins of Life* was one where the primitive atmosphere was reducing, with no free oxygen, and was acted upon mainly by ultraviolet radiation and electric discharges. Oparin's ideas remained untested for nearly 30 years until the early 1950s, when Harold C. Urey at the University of Chicago, having read Oparin's book, decided to investigate the primitive atmosphere from a physical–chemical point of view. Urey suggested that the carbon, nitrogen, oxygen and sulfur on the primitive Earth were present in their most reduced forms: carbon as methane (CH_4), nitrogen as ammonia (NH_3), oxygen as water (H_2O), and sulfur as hydrogen sulfide (H_2S). Urey considered carbon dioxide (CO_2) to have been a minor constituent of the primitive atmosphere, but thought there was probably some hydrogen. A student of Urey's, Stanley L. Miller, created Urey's primitive atmosphere in a sealed glass apparatus consisting of two connected flasks. The apparatus is illustrated in Figure 4.8. He half-filled the lower flask with water (to represent the ocean), then after evacuating the air from the apparatus, he added to the upper flask 1 part of hydrogen and 2 parts each of methane and of ammonia (to represent the primitive atmosphere). The water was boiled in the lower flask; the water vapour then passed into the upper flask, where it mixed with the 'atmosphere'. For about a week electric discharges were passed through this mixture of gases (to simulate lightning). At the end of the experiment the upper flask was stained a deep red colour due to the synthesis of organic molecules. These included some of the 20 amino acids that occur in proteins, together with many that do not, but that have nonetheless been found in meteorites (see Table 4.1). Among the other products found were: fatty acids, hydroxy aldehydes and sugars; purines such as adenine and guanine, as well as the pyrimidine thymine. As we saw in Section 4.3, these products are among the basic constituents of DNA and RNA and of the A–T G–C code common to nearly all life on Earth. Thus Oparin's ideas on the origin of organic molecules and their possible pre-biotic (pre-biogenic) synthesis were successfully put to the test for the first time. In the light of Crick and Watson's discoveries, Miller and Urey's results looked very promising indeed (and, in fact, recent experiments by Miller have also produced the other pyrimidines, cytosine and uracil). However, there is one major problem – the atmosphere used by Miller and Urey is not the one now thought to have existed on the primitive Earth.

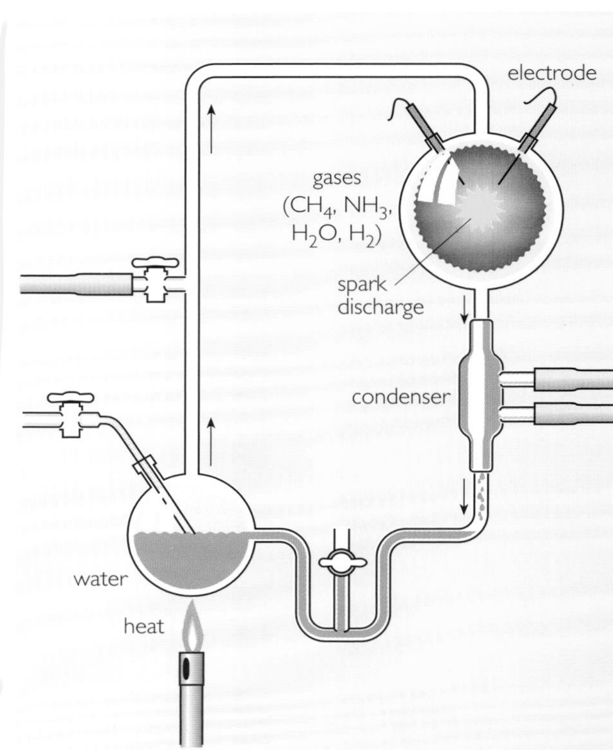

Figure 4.8
Miller and Urey's apparatus used to synthesize amino acids. The lower flask, half-filled with water, represents the 'oceans', the upper flask contains gases to represent the 'atmosphere'.

Table 4.1 Some of the amino acids synthesized in the Miller–Urey experiments which are also present in the Murchison carbonaceous chondrite (see Section 4.4.3) and are found in proteins on Earth. You are *not* expected to memorize the names of these amino acids.

Amino acid	Synthesized in the Miller–Urey experiments	Found in proteins on Earth	Found in the Murchison meteorite
glycine	✓	✓	✓
alanine	✓	✓	✓
α-amino-*N*-butyric acid	✓		✓
α-aminoisobutyric acid	✓		✓
valine	✓	✓	✓
norvaline	✓		✓
isovaline	✓		✓
proline	✓	✓	✓
pipecolic acid	✓		✓
aspartic acid	✓	✓	✓
glutamic acid	✓	✓	✓
β-alanine	✓		✓
β-amino-*N*-butyric acid	✓		✓
β-aminoisobutyric acid	✓		✓
γ-aminobutyric acid	✓		✓
sarcosine	✓		✓
N-ethylglycine	✓		✓
N-methylalanine	✓		✓

- What might have been wrong with the chemistry of Miller and Urey's atmosphere?

- As you read in Section 3.6.1, the Hadean atmosphere is now thought to have consisted of water vapour, carbon dioxide, nitrogen and sulfur oxides. In other words, it is now thought that the atmosphere in which life emerged was only mildly reducing or even slightly oxidizing.

One reason why our ideas about the Hadean atmosphere have changed since the time of the Miller–Urey experiments is that our ideas about the origin of the Moon have changed. In Section 3.4 we described how the Moon is now believed to have separated from the Earth about 4.5 billion years ago, i.e. very soon after formation of the Earth, but *also* after the Earth had acquired a semblance of the layered structure we recognize today (Figure 3.2). Before this happened, the Earth was probably more-or-less homogeneous (the early 'fruit-cake mix' Earth described in Section 3.3). If the molecules listed in Tables 2.1 and 2.2 are any guide, any primitive atmosphere on this early Earth would have consisted largely of *reduced* molecules originating from the accreting disc of the solar nebula (Section 2.4.4) and outgassed from the 'fruit-cake mix' Earth as a consequence of heating by continued impacts during and after accretion to form the planet. This atmosphere may have contained substantial concentrations of hydrogen (H_2), with carbon occurring as carbon monoxide (CO) and methane (CH_4), nitrogen as both N_2 and ammonia (NH_3), and possibly sulfur as hydrogen sulfide (H_2S). The cataclysmic upheaval associated with formation of the Earth's iron-rich core, coupled with the event which formed the Moon (Section 3.4), can hardly have failed to blast this primitive atmosphere of reduced molecules into space, i.e. so much heat was generated during the collison that the gases simply escaped the Earth's gravitational field (Section 3.6.1). Once the bulk of the Earth's metallic iron had been removed to the core, leaving a relatively more oxidized mantle consisting largely of silicates, the principal products of outgassing are likely to have been more oxidized and dominated by carbon dioxide (CO_2), nitrogen (N_2) and possibly sulfur dioxide (SO_2).

This may be a consensus view about the nature of the Hadean atmosphere but it is not necessarily accepted by all scientists. Research in the years since Miller and Urey's classic experiments has explored a range of plausible models, from those with a strongly reducing atmosphere to those with a more oxidizing one.

We will digress briefly here to comment on the red colour of the organic products of the Miller–Urey experiments. You may recall that Mars is known as the Red Planet (Section 1.1.1) and that its red colour is due to iron oxides, *not* organic molecules. This was in fact known before the Viking missions to Mars, partly because an atmosphere dominated by CO_2 is incompatible with a surface coating of organic molecules, but chiefly because the spectral 'signature' of light reflected from the martian surface matches that of iron oxides and not of organic molecules.

However, this does not mean that *all* red objects in the Solar System owe their colour to iron oxides. Recent observations of the Kuiper belt (Section 2.4.5) have identified a total of 17 objects with diameters between about 100 and 300 km, whose nature and origin remains enigmatic, for they seem to be neither comets nor asteroids. One of these bodies, named Pholus, is said to be the 'reddest' object in the Solar System and this is almost certainly because its surface is indeed covered with organic materials. (See also Section 4.4.3.)

We must bear in mind that the outer planets of the Solar System probably still have most of their primordial atmospheres composed of reduced gases – an obvious example is Jupiter (Section 2.3.2) which is known to have an atmosphere rich in hydrogen, methane and ammonia, along with water. We may conclude that while the Miller–Urey experiments may not have mimicked conditions on the early Earth, they may have approximated quite well to reactions that occur on the surfaces of other planets and bodies further out in the Solar System.

Later experiments

In the decades following the Miller–Urey experiments, some geochemists favoured a model for the atmosphere based on the range of gases emitted by modern volcanoes and dominated by carbon dioxide, nitrogen and water vapour (cf. Section 3.6.1). Others argued that volatile materials from outgassing had interacted with the Earth's crust to produce an atmosphere consisting of carbon monoxide, carbon dioxide, nitrogen and hydrogen. Both these atmospheric models are less reducing than Urey's and further experiments examined other possible pathways for the abiotic synthesis of organic molecules. Various combinations of gases were tried:

1 CO, N_2, H_2 – Electric discharges acting on this mixture were not particularly effective at producing amino acids. However, methanal (formaldehyde, CH_2O), the simplest molecule necessary for the pre-biotic synthesis of complex sugars, was produced in large quantities. The sugars ribose and deoxyribose are critical constituents of RNA and DNA, as they provide the links to the nucleotide bases (Section 4.3).

2 CO_2, N_2, H_2 – Here too, electric discharges produced only low yields of amino acids.

3 CO, H_2 – This mixture is used commercially to produce hydrocarbons by a process called the *Fischer–Tropsch reaction*. The reaction requires a catalyst, usually iron or nickel and a temperature of 200–400 °C. Depending on the reaction conditions, a variety of different organic compounds can be produced. If ammonia (NH_3) is added to the CO and H_2, then amino acids can be formed, cf. item 2 above.

- What is the constituent common to all three combinations?
- Hydrogen, which does not accord with most modern ideas about the atmosphere of the early Earth.

Another problem with hydrogen as an atmospheric constituent is that, because of its very low relative atomic mass, it escapes from the Earth's gravitational field and is rapidly lost to space, i.e. it has a very short *residence time* in the atmosphere.

If the early Earth's atmosphere was dominated by carbon dioxide, nitrogen and water, and if hydrogen and other reduced species were in short supply or unavailable, how might the molecular building blocks of life have been produced?

One possibility is mediation by inorganic minerals. In 1960, scientists irradiated a solution containing reduced iron(II) sulfate and carbon dioxide with light and got a yield of methanal (formaldehyde, CH_2O) of approximately 1%. This suggests that dissolved iron(II) might have provided an important source of reducing power on the early Earth. Iron is one of the most abundant elements in the Earth's crust, particularly in the magmas that rise from the mantle to form basaltic oceanic crust (Figure 3.3). Fe^{2+} ions could have entered the surface environment at hydrothermal vents (Figures 3.16 and 3.17) associated with submarine volcanic activity (see Section 4.4.2). Other experiments have shown that ultraviolet irradiation of water containing suspended clay particles and dissolved CO_2 can result in the production of methanol (methyl alcohol, CH_3OH). Clay minerals would have been readily available on the early Earth from the weathering of rocks exposed to the atmosphere. Clay minerals are silicates with a layered structure that can be very complex and can carry surface charges enabling them to attract and capture (**adsorb**) other molecules. Storms and meteorite impacts would have lifted the clays into the CO_2-rich atmosphere to be exposed to ultraviolet radiation and so catalyse the formation of complex organic molecules. Even if *formed* in the atmosphere, however, the first organic molecules must have ended up in the oceans.

4.4.2 Hydrothermal vents

The possibility of continued intense bombardment by meteorites long after the Earth had formed (Section 3.5) has led some scientists to conclude that the emergence of life anywhere near the surface of the oceans was not possible until perhaps as late as 3.8 billion years ago. This in turn led some investigators to suspect that life did not originate in 'warm little ponds' at the Earth's surface, as had long been proposed, but in the ocean depths.

Discovery in the late 1970s of hydrothermal vents on ocean ridges (Figures 3.16 and 3.17) coupled with the fact that among the simplest modern organisms, bacteria, are thermophilic varieties, led some researchers to suggest that life could have originated at hydrothermal vents, which must have been abundant in the Hadean ocean (Section 3.6.1).

- Apart from the very different temperature regimes, there is one rather important difference between the environment of 'warm little ponds' on the one hand and hydrothermal vents on the other. Bearing in mind the different settings, can you suggest what it might be?

- The oceans are dark below depths of a few hundred metres and it would be impossible to synthesize even the simplest organic molecules by using the energy from solar radiation, because it would all (especially the ultraviolet) have been absorbed by the water.

Nonetheless, as numerous glossy pictures in popular science magazines bear eloquent witness, hydrothermal vents support highly diverse communities in the deep sea at the present time and the notion that they might be the 'cradle of life' has a certain appeal for some scientists. However, as light is not available as a primary energy source for these organisms, the primary producers in these communities are autotrophic bacteria (cf. Section 3.2.1). These obtain their energy by the oxidation of reduced sulfur compounds that are supplied in the hot vent waters (the primary energy is *not* provided by the hot water). In simplified form the reaction is:

$$\underbrace{CO_2 + H_2S + O_2 + H_2O}_{\text{in solution}} \xrightarrow{\text{chemical energy}} \underbrace{CH_2O}_{\substack{\text{organic}\\\text{matter}}} + \underbrace{2H^+ + SO_4^{2-}}_{\text{in solution}} \qquad \text{(Equation 4.1)}$$

where CH_2O represents the 'building blocks' of carbohydrates $(CH_2O)_n$.

This process is called **chemosynthesis** and the reaction differs from that of photosynthesis (Reaction 1.4, Box 1.4) in two important ways:

- First, oxygen is not liberated, but *used* in the reaction. It oxidizes the hydrogen sulfide (H_2S) and so provides the chemical energy for synthesis of carbohydrate, $(CH_2O)_n$.
- Second, the hydrogen used to reduce the CO_2 comes from oxidation of the hydrogen sulfide, not from water. The water combines with SO_2 produced by sulfide oxidation, to produce hydrogen ions (H^+) and sulfate ions (SO_4^{2-}).

- Why would Reaction 4.1 have been impossible in the Hadean ocean?

- With an oxygen-deficient atmosphere, there could have been no free oxygen in solution in seawater. The oceans were therefore anoxic.

Any chemosynthetic bacteria present in these anoxic oceans would have been anaerobic varieties. In these circumstances the only means of acquiring chemical energy to liberate hydrogen from H_2S would have involved the concomitant reduction of oxidized chemical species like sulfate ions (SO_4^{2-}) or nitrate ions (NO_3^-), which would almost certainly be far too expensive in energy terms to be a viable proposition.

However, if such environments were near the surface, then anaerobic bacterial *photosynthesis* could have occurred, using sulfide (H_2S) to provide the necessary hydrogen as before, with solar radiation as the energy source:

$$CO_2 + 2H_2S \xrightarrow{\text{solar radiation}} CH_2O + 2S + H_2O \qquad \text{(Equation 4.2)}$$

- Do we know of any present-day examples of such bacteria?

- Yes. There are two groups of anoxygenic photosynthetic bacteria: the green and purple sulfur bacteria (Figure 3.7b).

There are also plenty of modern examples of the most primitive types of bacteria, the Archebacteria or Archea (Box 4.3), which are *hyperthermophilic* (cf. Section 3.5). They live in and near hot springs, volcanic fumaroles (steam vents) and hydrothermal vents at temperatures of between 90 and 110 °C. It is possible that the earliest organisms may also have been hyperthermophilic bacteria, which means they would have thrived in the hot waters round hot springs and/or hydrothermal vents, where concentrations of hydrogen sulfide would also have been greatest. Although modern chemosynthetic bacteria do not depend on the heat from hydrothermal vents as an energy source, at least some purple and green sulfur bacteria make use of light from both ends of the visible spectrum for their (anoxygenic) photosynthesis. Could the earliest archebacteria have used thermal (infrared) radiation from hydrothermal vents for their photosynthesis? It seems that it might just be possible, provided that the bacteria really were thermophilic and able to operate in water of 50–100 °C (i.e. within a few tens of centimetres of the main plume, Figure 3.17). If that were the case, then it is not too implausible to envisage the deep oceans as a potential 'cradle of life'. However, the balance of probability would seem to favour a setting nearer the surface, not least because solar radiation – even from the relatively weak Sun of Hadean times – would have been a great deal more intense source of energy than thermal radiation from hydrothermal vents.

In summary then, we can envisage either relatively near-surface hydrothermal vents or hot springs on volcanic islands, populated by anoxygenic photosynthesizing thermophilic archebacteria; these would be synthesizing organic molecules from dissolved carbon dioxide via dissociation of the abundant hydrogen sulfide emitted by the vents, using solar radiation as the energy source. It is more difficult to envisage how and when such organisms might have evolved into oxygenic autotrophs and we return to this topic later. In the meantime, it is noteworthy that in the mid-1990s this route for the origin of life on Earth was gaining increasing support among scientists as one of the most plausible.

Question 4.5

(a) Which of the experiments summarized in Section 4.4.1 produced the greatest yield of amino acids and other *complex* molecules?

(b) If simple molecules such as methanal (formaldehyde) and methanol (methyl alcohol) are the principal products of a laboratory experiment with the constituents of a relatively oxidized atmosphere, how is it possible to surmise that the basic building blocks of life might not have formed on Earth at all?

4.4.3 Space invaders

We have already suggested that asteroid and comet impacts may have prevented life on Earth from getting started for many millions of years, or snuffing it out just when things seemed set for a rosy future. However, asteroids and comets may also have performed a crucial role in providing many of the essential building blocks for life. Moreover, carbon-rich meteorites enable us to test the proposition that the organic compounds necessary for life can form naturally by non-biological (abiotic) processes.

Tables 2.1 and 2.2 show that there is a great variety of molecules in space. They range in complexity from simple molecules with only two atoms (e.g. OH, CO, CN) to compounds such as hydrocarbons, amines, alcohols and nitrites.

An interstellar molecular cloud with temperatures as low as –170 °C or less and as few as a billion particles per cubic metre (Section 2.4.2), may seem an unlikely place for chemical reactions to occur. Yet the variety of complex organic molecules that are present in the outer Solar System, in comets that come from far beyond the outermost planets and in interstellar molecular clouds, is strong evidence that the crucial 'raw materials' of life are widespread throughout our galaxy. However, a bone-dry particle of interstellar dust baked by ultraviolet light and cosmic rays is an unlikely place for the *origin* of life, even if it does carry organic molecules. Having said that, there are bacteria on Earth which can survive extremes of temperature, dessication and irradiation, though they have almost certainly evolved from forms that originated under more benign conditions. Nearly all life seems to require liquid water, which in turn needs a planet, because in the near vacuum of space water can exist only as a gas or as a solid (ice).

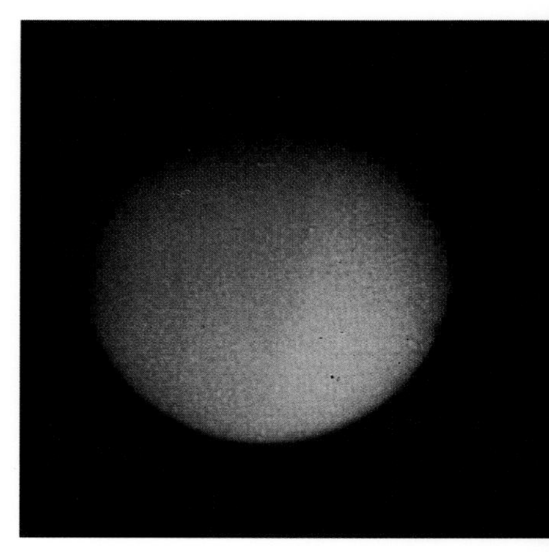

Figure 4.9
Voyager image of the largest moon of Saturn, Titan, in visible light. The reddish orange colour is due to the presence of organic compounds in Titan's atmosphere.

There is quite a lot of water in our solar system and organic compounds appear to be ubiquitous. On Jupiter (Figure 3.22), the presence of simple hydrocarbons in the atmosphere gives the colours to its clouds and is a strong indicator of active organic chemistry. The presence of the hydrocarbons could readily be explained as the products of ultraviolet photochemistry and lightning activity. Harold Urey was the first to suggest that the colour of Jupiter's atmosphere was due to organic molecules. However, perhaps one of the most interesting worlds is Saturn's giant moon, Titan (Figure 4.9). When Voyager 2 approached Titan in 1981 scientists discovered that they could not see the surface because it is obscured by an opaque, reddish orange haze. Voyager's instruments showed that Titan has an atmosphere 10 times denser than the Earth's, even though Titan is smaller (about the size of Mercury). Its atmosphere is composed mainly of nitrogen but contains up to 10% methane and other simple organic molecules. This is plainly a reducing environment, and as we saw in Section 4.4.1, electric discharges in a reducing atmosphere can produce a range of (reddish) organic compounds.

Ultraviolet light from the Sun, charged particles and cosmic rays all bombard Titan's atmosphere and can initiate chemical reactions. Carl Sagan and his colleagues at Cornell University simulated Titan's atmosphere in the laboratory and irradiated the gases with charged particles. The experiments produced dark, tar-like organic solids that they called *tholins*, which have almost identical spectral characteristics to the reddish haze in Titan's atmosphere. Sagan suggested that organic molecules continually form in the upper atmosphere of Titan and slowly fall to the surface as tholins. If this process has continued over the past four billion years, Titan's surface must be covered by a layer of organic compounds, tens to hundreds of metres thick. This raises the prospect of some interesting chemical reactions should Titan's surface ever become warm enough to melt any of the ice which is locked up in its crust. When Sagan and his colleagues mixed tholins with water in the laboratory they made amino acids and also discovered traces of nucleotides. Moreover, although Titan's surface temperature is normally –179 °C, they estimated, from comparisons with other satellites, that Titan has had around a 50% chance of having experienced periods in its history, perhaps hundreds of years long, when liquid water was released by the heat from impacts. Is 1000 years long enough for primitive life-forms to have developed on Titan following an impact, when lakes of water–ice slurries might briefly have formed? Watch this space: the first close-up look at the surface of Titan is scheduled to occur when the ESA–NASA Cassini mission arrives in the Saturn system around 2004 and launches the Huygens probe into Titan's atmosphere.

One of Jupiter's moons, Europa (which is nearly as big as our own Moon), also has a crust of ice over its surface and this crust is not cratered, which implies a 'self-healing' process after the inevitable puncturing by impacts. This suggests that enough heat may come from inside Europa to maintain at least some pockets of liquid water beneath the ice crust.

Organic compounds in meteorites

Carbonaceous chondrites provide us with the most ancient organic molecules available for study (Section 2.3.2). If these compounds were formed in the Solar System they could be as much as 4.6 billion years old, but if they derive from interstellar molecular clouds they could be even older. Today the laboratories and museums of the world have some 25 000 meteorite samples, a relatively small proportion of which (about 5%) are carbonaceous chondrites. Of these, one of the most celebrated is the Orgueil meteorite, which fell in the south of France in 1864 (Figure 2.10). About 20 fragments were collected totalling some 12 kg and it was one of those fragments which was used to perpetrate the cosmic 'Piltdown hoax' described in Box 4.2. Another well-documented carbonaceous chondrite arrived more recently. At 11am on Sunday 28 September 1969, the appearance of a fireball in the skies above eastern Australia was followed by a thunderous explosion. A total of some 500 kg of meteorite fragments fell near the town of Murchison, 85 miles north of Melbourne. The first person on the scene reported a smell like 'meths' (methanol), strongly suggestive of organic compounds.

The presence of organic material in carbonaceous chondrites was first reported in 1834 by the Swedish chemist, Jons Jakob Berzelius. He found that by extracting a sample with water and then using distillation techniques he obtained complex organic substances. Berzelius's work 160 years ago began the intensive study of organic compounds in meteorites which continues today.

During the 1950s and 1960s scientists identified a variety of organic compounds in meteorites including amino acids. However, the ever-present risk of contamination by terrestrial organic materials meant that many of the reports of organic compounds in meteorites were regarded with scepticism. That applied not only to the possible life-forms discussed (and dismissed) in Section 4.2.1, but to the various amino acids and other molecules as well.

The Apollo programme in the United States during the 1960s and early 1970s led to the development of a vast array of sophisticated analytical techniques designed to look for organic compounds in the lunar samples. So when the Murchison carbonaceous chondrite fell only months after the return of the first lunar samples, scientists turned this battery of analytical equipment to its study. The precise methods of gas chromatography combined with mass spectrometry gave unequivocal evidence for the presence of organic compounds in the Murchison meteorite. Moreover, biologically produced amino acids can be distinguished from those formed abiogenically by their *chiral* properties, as described in Box 4.5. In brief, nearly all amino acids in living organisms are of the L-form, whereas those in the Murchison meteorite were present as both forms, the left-handed and the right-handed, in equal amounts. This could not be due to contamination because careful study had identified amino acids that are not found in proteins on Earth. This does not provide confirmation of cosmic credentials, however, because many hundreds of non-protein amino acids have been identified in plants. Truly cosmic credentials would be provided by one or more amino acids in the Murchison meteorite that are *not* found in any living organism – but none has yet been identified.

Box 4.5 Amino acids and chirality

$$\begin{array}{c} H \\ | \\ NH_2-\mathbf{C}-COOH \\ | \\ H \end{array}$$

glycine

$$\begin{array}{c} H \\ | \\ NH_2-\mathbf{C}-COOH \\ | \\ CH_3 \end{array}$$

alanine

$$\begin{array}{c} H \\ | \\ NH_2-\mathbf{C}-COOH \\ | \\ CH_2 \\ | \\ Ph \end{array}$$

phenylalanine

$$\begin{array}{c} H \\ | \\ NH_2-\mathbf{C}-COOH \\ | \\ CH_2 \\ | \\ OH \end{array}$$

serine

(a)

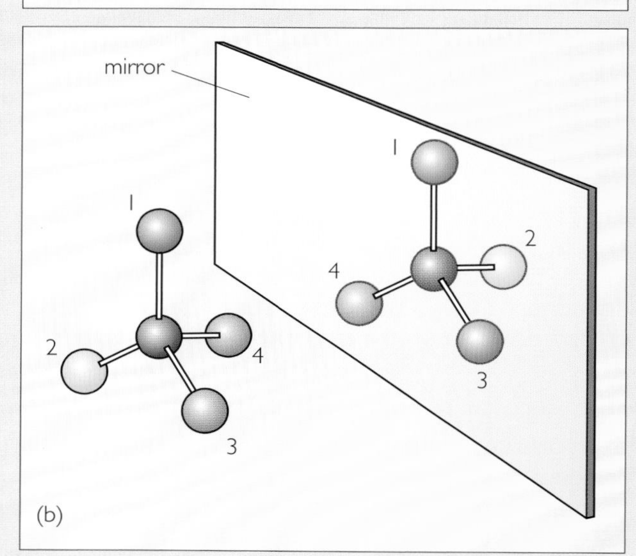

Figure 4.10
(a) Two-dimensional representation of the structures of some common amino acids. The 'Ph' in phenylalanine is a phenyl group (a ring made up of six carbon atoms). The chiral centres in the four amino acids are shown in bold. (b) Illustration of chirality. The two molecules are mirror-images of each other.

Amino acids contain carbon, hydrogen, nitrogen, oxygen and, on occasion, sulfur atoms. There are about 20 different kinds of amino acid involved in the structure of most kinds of protein molecule (Box 1.5). The formulae and names of four of the most common amino acids are shown in Figure 4.10a. You do *not* need to memorize them.

Each amino acid has three common structural units, an amino group ($-NH_2$), a carboxyl group ($-COOH$), and a hydrogen ($-H$). The fourth group can be anything from a simple hydrogen atom (in glycine) to a more complex unit containing a carbon ring (in the case of phenylalanine) and it is this fourth group that distinguishes amino acids from one another. Amino acids also exist as **isomers**, i.e. compounds that have the same molecular formula but with the atoms combined in a different order. In fact, amino acids can form special kinds of isomers that have profound consequences for the chemistry of life. The key to this type of isomer is the three-dimensional arrangement of atoms or groups around a central carbon atom called a *chiral* centre. The ability of the carbon atom to form bonds with four other atoms or groups means it can form molecules that have the shape of a tetrahedron, the simplest example being methane, as we saw in Figure 1.5a. When the four atoms or groups are different from each other it is possible for the molecule to have two forms, each a mirror image of the other. This phenomenon is called **chirality**. Figure 4.10b shows such a molecule represented as a 'ball and stick' model with the amino, carboxyl and hydrogen groups of the amino acid represented by different coloured balls. You can see that no matter how it is twisted or rotated the two forms of the molecule cannot be superimposed on each other. You can see this for yourself by trying to superimpose your left hand on your right: instead of atoms, it is your fingers that are connected in a different order and your left hand is a mirror image of your right.

The majority of amino acids in living organisms are found in only one of these two mirror-image or chiral forms, called by convention the L-form. The other is called the D-form and is the one typically produced by laboratory synthesis of amino acids. The convention derives from the Latin words *dexter* meaning 'right' and *laevus* meaning 'left' and the molecules are said to be left- or right-handed, or negative (−) and positive (+), respectively.

As you might expect, chirality is not confined to amino acids but it remains true that most chiral molecules made by living organisms are found in only one of the two possible forms, usually the L-form (but not always, see below).

Moreover, the synthesized or unnatural form usually behaves differently from the natural form when introduced artificially into a biological environment. For instance, of the two isomers of glutamic acid, only the naturally occurring form (D-form, in this case) functions as an effective flavour-enhancing agent for meats, soups and other foods. This whole question raises a fascinating point on which to speculate. How did it come about that we live in a world in which, for instance, both natural alanine (Figure 4.10a) and natural glucose are 'right-handed'; whereas for many other organic compounds, the naturally occurring form is the 'left-handed' one? If all the naturally occurring isomers of chiral molecules could be simultaneously inverted to the other form would we be any different? By definition, isomers are chemically indistinguishable and the two isomers of a chiral molecule can only be distinguished by X-rays or by interaction with polarized light; but generally only one of a pair of isomers is of biochemical importance.

Nevertheless, the chemical processes thought to have given rise to life are not some exotic improbable set of reactions, but the normal chemistry of carbon, oxygen, hydrogen, nitrogen and sulfur throughout the Universe. However, pathways for the chemical development of organic matter towards the evolution of life are governed by the conditions of a particular environment, be it an interstellar dust particle, a meteorite or a planet. For example, the planets, Venus, Earth and Mars all experienced a similar early history but only on Earth did life persist and evolve. We cannot rule out the possibility that life arose on Venus and Mars also, but did not survive because of later changes in their environments (but see Section 4.6).

Could extraterrestrial organic molecules have been a significant or even a major source of building blocks for pre-biotic synthesis on the Earth? As we saw in Section 3.3, large impacts deliver such stupendous amounts of energy that upon impact they would destroy any organic matter they contained. Many smaller meteorites simply 'burn up' in the atmosphere, but some do not, otherwise there would be no meteorites to collect. Indeed, extraterrestrial sources are delivering organic molecules to the Earth all the time, in interplanetary dust particles. Most of these are thought to be derived from comets small enough to be gently decelerated by the atmosphere (Figure 2.19) and from meteorites or meteorite fragments large enough to avoid complete 'burn up', but small enough to be substantially decelerated during their passage through the atmosphere. In particular, carbonaceous chondrites cannot have been strongly heated, because if they had been, no organic compounds would have survived. Such supplies of organic matter must have been available to the Hadean Earth.

Question 4.6

It is estimated that some 3×10^5 tonnes (3×10^8 kg) of cosmic dust falls on the Earth each year (Section 2.4.5). If 5% of that is from carbonaceous chondrites and if 3% of that is made up of organic molecules, what is the annual flux of organic 'space invaders' to the Earth's surface, given a surface area of 5×10^{14} m^2? Your answer will be in kg m^{-2} yr^{-1} (kilograms per square metre per year).

The answer to Question 4.6 does not seem much, but even a complex organic molecule with a relative molecular mass in the thousands (10^3 to 10^4) will have between about $10^{23}/10^3$ and $10^{23}/10^4$ molecules per gram, i.e. between 10^{22} to 10^{23} molecules per kilogram. (If you are unsure where these figures come from, look back to Box 2.1 where a similar calculation was carried out.) So a billionth of a kilogram (10^{-9} kg, the answer to Question 4.6) would still have about 10^{13} to 10^{14} molecules even if the flux were perfectly uniformly distributed, which is very unlikely to be the case.

- Would you expect this 'space invader' flux to the Hadean Earth to have been greater or less than it is now?

- It would almost certainly have been considerably greater if our inferences in Section 3.3 are correct.

4.5 Making larger molecules

We have seen that under the right conditions, it is relatively simple to make organic molecules in the laboratory over a period of a few weeks. Given the many millions of years that were available early in the history of the Solar System, much more than this simple chemistry would have taken place. In addition to this, scientists' test-tubes are clean and filled with only a few chemicals, compared with the great mixtures of elements and compounds and highly variable conditions that would have characterized early planetary environments. In other words, abiogenic synthesis of organic molecules is no great mystery; we merely have no record of the pathways/reactions by which they were formed in the Hadean.

The basic necessities for life are self-replication, energy and chemical sustenance, and a membrane to hold everything together. As we saw in Section 4.3, all life on Earth is based on RNA, DNA, or both, and DNA molecules contain the four nucleotide bases adenine, thyamine, guanine and cytosine — the A–T G–C code common to all organisms with the exception of viruses which only contain RNA. How did nature achieve all this and how did it arise? One central question is whether this A–T G–C instruction sheet for life was the only one to have evolved. Perhaps others also arose on the early Earth and this is the only survivor, or the one that was most successful. Another question is what came first, the A–T G–C code or life? Did this 'model' for life arise first and then build life-forms around it, or did some precursor chemistry capable of self-replication develop which then acquired a membrane that enabled the template for life (A–T G–C code) to be developed inside it? Various hypotheses have been proposed to answer the question of how membranes originated. One suggestion is based on the fact that some organic molecules are polarized, that is, they have a polar structure: one end has an affinity for water, the other end repels water — one example is an alkylamine with the formula $C_{10}H_{21}NH_2$. Some of these polar molecules, called **amphiphiles**, have a remarkable property: they assemble into stable *lamellar* structures, i.e. structures which are made up of layers of molecules. If these molecules are left in contact with water they have a tendency to curl up and make tiny cell-like spheres. It has been suggested that some of the organic compounds present in meteorites display amphiphilic properties and that extraterrestrial material could have been a significant source of the first membranes that may have gone on to form cells.

Many scientists believe that once some form of protocell or membrane was available to encapsulate the machinery of life (its metabolism) and segregate it from its surroundings, life was well on its way. However, while such membrane structures may be essential to the origin of life, they do not themselves reproduce or evolve.

If we suppose that simple molecules such as amino acids were the major products of early pre-biotic synthesis and that they were in the oceans and so protected from ultraviolet light, the next question is how did they reach sufficiently high concentrations to combine into more complex, stable compounds? Compared even with the complexity of proteins, the pre-biotic route to nucleic acids is very difficult. Some essential constituents of nucleic acids such as purines (Section 4.3), including adenine and guanine can be derived quite readily from hydrogen cyanide (HCN), but it is much harder to form pyrimidines (thymine, cytosine and uracil) and there is no obvious pre-biotic route – although all were formed in the Miller–Urey experiments (Section 4.4.1). The conversion of purine or pyrimidine bases into nucleotides (with sugar and phosphate group components, Figure 4.2), and the polymerization of these units to give nucleic acids capable of self-replication is even more difficult to envisage. Reviewing the problems in 1989, Gerald Joyce concluded that they were insuperable and that life did not start with nucleic acids: they came later after a simpler replicating system had somehow developed.

One possible mechanism for development of such a system, proposed in the 1950s by the physicist Joseph Bernal, extends the role of clays (mentioned at the end of Section 4.4.1), which were probably abundant on the early Earth. In brief, Bernal suggested that the electrically charged surfaces of clay particles, acting rather like catalysts, could adsorb, concentrate and organize organic molecules to form replicating templates. Quite how nucleic acids were synthesized on clay surfaces and eventually took over the replicating template role and how cells evolved from such clay 'proto-life,' remain unresolved questions. Similar problems beset an alternative suggestion, that the iron sulfide mineral, pyrite (fool's gold) FeS_2, might have provided charged surfaces capable of attracting organic molecules and catalysing their organization into more complex molecular structures.

4.5.1 Origin of genetics – chicken or egg?

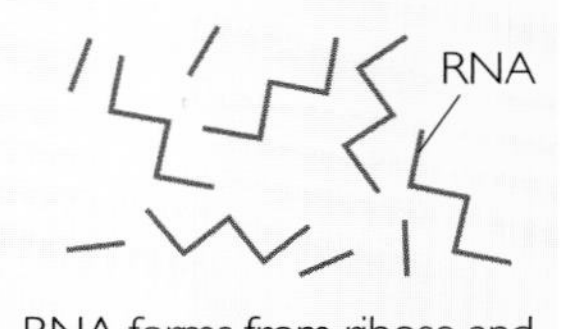

RNA forms from ribose and other organic compounds.

As RNA molecules evolve, they 'learn' to copy themselves.

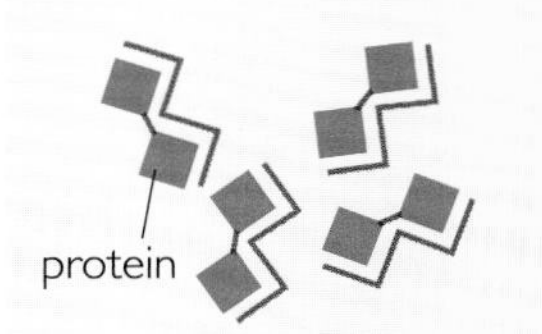

RNA molecules begin to synthesize proteins that can serve as catalysts.

Figure 4.11
The early stages in a model for the evolution of an RNA world.

In nearly all living organisms today, the genetic machinery consists of self-replicating DNA molecules. It is an elegant and highly complex system, but how did it arise? Proteins are themselves the products of the translation process and their synthesis in turn depends on the presence of pre-formed nucleic acids. The origin of the genetic process thus might seem to be a chicken-and-egg problem; which came first – proteins or the coded nucleic acids? Whether we start with replicating clays, pyrite systems or something completely different, nucleic acids must have evolved at some very early stage and taken on their role in replication and translation.

There are now good reasons for believing that RNA, and not DNA, was the first nucleic acid to perform this role. The evidence came in a discovery made independently by the US biochemists Sidney Altman and Thomas Cech which led to them being jointly awarded the Nobel Prize for Chemistry in 1989. They found that certain types of RNA can have catalytic activity, similar to that of protein enzymes, snipping themselves in two and splicing themselves back together again. So RNA could serve both as gene and catalyst, egg and chicken, and replicate itself without the help of proteins. This led to the idea that RNA was the genetic material of early cells (protocells) and may once have been involved in far more cellular activities than it is now.

Even under the best conditions, RNA and its components are difficult to synthesize. For example, the process by which the sugar ribose, a key part of RNA, is made also produces a large number of other sugars that would actually *prevent* RNA synthesis. An outline of how an RNA world might have arisen is as follows (Figure 4.11). Once RNA has been synthesized from ribose and other organic compounds, it eventually becomes self-replicating. RNA molecules then begin to synthesize proteins that act as catalysts and in turn allow the RNA to synthesize more proteins. These proteins eventually help the RNA to make double-strand copies of itself and so evolve into DNA. However, before the transition from the RNA world to the DNA world occurred, the complex machinery for translating nucleic acids into protein sequences must also have evolved. At some stage during evolution, DNA must have taken over from RNA. The DNA must have used the RNA to make the proteins which in turn help DNA to make copies of itself and transfer the genetic information to RNA. Today all living organisms use DNA as their genetic material, with the exception of RNA viruses, some of which may be molecular fossils of the RNA world.

4.6 Microbes from Mars

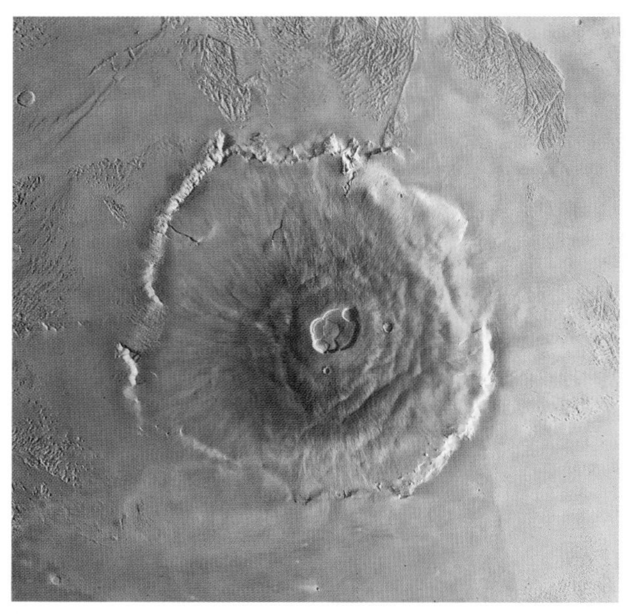

Figure 4.12
Olympus Mons – the youngest, and largest, volcano on Mars. The summit is about 24 km above its base and the crater is about 60 km in diameter and up to 3 km deep.

We left Mars in Chapter 1 with the negative results of the Viking lander experiments, which revealed no extant life on Mars (at least within the surface layers at the landing sites), but did not exclude the possibility of ancient long-extinct life. New evidence accumulating in the succeeding 20 years gave new impetus to this possibility. Images of Mars from spacecraft missions such as Mariner and Viking led scientists to conclude that many of the surface features on the planet are consistent with their formation by running water. It is possible that before about 3.8 billion years ago the planet had a cool climate and abundant supplies of groundwater, perhaps mostly in frozen form, but melted and forced to the surface by geothermal and impact-related heat sources. It seems unlikely that the carbon dioxide-rich atmosphere would have been able to produce sufficient greenhouse warming to offset the relatively weak radiation from the early Sun (25–30% less than now); so there was no rain, nor any permanent river system, let alone an ocean. Lakes possibly favourable for the development of life could have formed in impact craters more than 10 km or so across, warmed by thermal energy from rocks melted by the impacts. Moreover, Mars is covered with extinct volcanoes (some of them enormous, see Figure 4.12), and it is further suggested that volcanic heat would have driven hydrothermal vent systems in these lakes – some of which could presumably also have formed in the volcanic craters themselves. In fact, some scientists have theorized that Mars *did* once have enough surface water for oceans to form (i.e. that the climate was warm rather than cool), in which case there would have been vastly more scope for the development of life.

A second strand of new evidence relating to possible ancient life on Mars comes from meteorites. The majority of meteorites originate in the asteroid belt (Box 2.3), but it has now become clear that a few have other sources. Analysis of the rocks point to an origin from both the Moon and Mars.

- How might that happen? How might fragments of the Moon or Mars reach the Earth as meteorites?

- Large impacts scatter fragments of the surrounding terrain over huge areas and to great heights. A tiny proportion of these are flung sufficiently high to escape the planet's gravitational pull and a tiny proportion of those are placed in Earth-crossing orbits.

In fact, by the mid-1990s there were about a dozen well-authenticated martian meteorites and a similar number of lunar ones – representing only about 0.1% of all meteorites collected (the lunar meteorites are fairly easily identified, using samples returned by Apollo missions for comparison).

Martian meteorites have presumably arrived on Earth throughout its history, but the recovered samples all appear to have taken between about 1 and 10 million years to reach us. However, that indicates only the time when they were ejected from the martian surface and it seems that the radiometric ages of the rocks themselves fall between 1 and 4.5 *billion* years. Examination of these meteorites has revealed evidence of precipitation of calcium and magnesium carbonates, presumably from aqueous fluids at or near the surface. At least one of the rocks contains complex organic compounds including polycyclic aromatic hydrocarbons (cf. Section 2.4.1), although there is no evidence of present-day biological activity in any of the martian meteorites.

From all this evidence, scientists now believe that it is at least possible that life could have originated on Mars early in its history, by one of the processes discussed earlier in this chapter. In 1995, some publicity was given to the possibility of hydrothermal vents as an environment in which martian life could have come into being, with headlines such as 'Search for White Worms on Mars' in tabloid newspapers.

- To what might such headlines be referring?

- If you look back to Figure 3.17b, you will see the white tube worms that characterize most hydrothermal vent communities in the Pacific (but not in the Atlantic; that is another story). The headlines imply that fossilized remains of similar worms might be found on Mars – which is rather unlikely (see Box 4.6).

Moreover, with the discovery of martian meteorites, the possibility that life originated on Mars and was then 'ejected' to Earth cannot be excluded. Descendants of that ancient life could even have survived to this day below the surface of Mars, either semi-dormant, like the bacteria reported from Antarctica, or as more active thermophilic bacteria in hot rocks at depth, like those reported from oilfields on Earth (cf. Section 3.5).

All in all, then, while we remain uncertain as to which of the routes to life reviewed in Sections 4.4.1 to 4.4.3 was actually followed, or if there was some other route or some combination of routes, we need no longer regard the Earth as the only place in the Solar System where life might have originated. Mars is one alternative. Another is the Jovian moon, Europa (Section 4.4.3), whose icy surface has been suggested by some scientists to be underlain by an 'ocean' up to perhaps 100 km deep, on the floor of which there could be life-supporting hydrothermal convective circulation systems driven by heat from the interior.

It is of course equally possible that life was dispersed the other way, that it originated on Earth and was then spread to other parts of the Solar System by terrestrial meteoroids ejected by the giant impacts that afflicted our planet up to about 3.8 billion years ago. It is no longer so implausible to accept space-travelling microbes, given what we know about the ability of some bacteria to remain dormant for millions of years and to withstand strong radiation and extremes of temperature (Section 4.2.1). Nonetheless, it would be unwise to suggest that even the most robust of microbial life-forms would be capable of surviving the kind of interstellar or intergalactic travel favoured by the advocates of panspermia.

Be all that as it may, however, we have seen the evidence from the rocks that there was life on Earth certainly 3.5 billion years ago and possibly 3.8 billion years ago. Irrespective of whether it originated here or came from somewhere else, it plainly strengthened its foothold and began to diversify, albeit slowly at first.

Those paragraphs were written early in 1996. A scant few months later, in early August, the following press release was flashed round the world:

> *A NASA research team of scientists at the Johnson Space Center and at Stanford University has found evidence that strongly suggests primitive life may have existed on Mars more than 3.6 billion years ago.*

In fact, the team also included scientists from McGill University (Canada) and the University of Georgia (USA). More details of this development are provided in Box 4.6.

Box 4.6 Fossils from Mars?

A two-year investigation by a NASA research team had by mid-1996 amassed an array of indirect evidence of past life on Mars. Inside an ancient martian rock that fell to Earth as a meteorite, the scientists found organic molecules, several mineral features characteristic of biological activity, and numerous tiny structures which are possible microscopic fossils of primitive bacteria-like organisms that may have lived on Mars more than 3.6 billion years ago. The 'mineral features characteristic of biological activity' include tiny crystals of iron sulfides and of magnetite (Fe_3O_4) both of which have long been known to be precipitated by terrestrial bacteria – though they are more commonly of inorganic origin. The largest of the possible fossils are of the order of 100 nanometres (nm, 10^{-9} m) in size (rather quaintly reported in the press release as 'less than 1/100th the diameter of a human hair'), most are ten times smaller (10 nm), which means that they are more comparable in size with terrestrial viruses than with bacteria. If you compare the sizes of these structures with those in fossil stromatolites (Figure 3.6), you will see that the martian 'fossils' are about three orders of magnitude smaller.

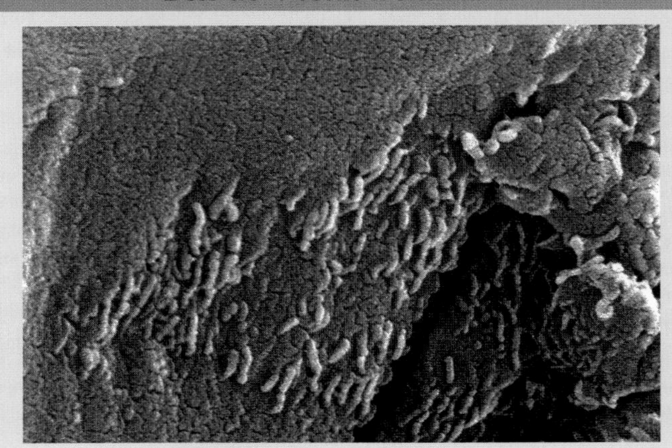

(a)

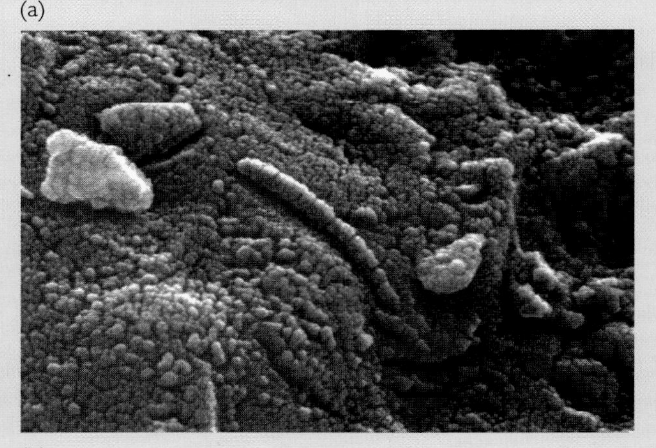

(b)

Figure 4.13
(a) and (b) Electron micrographs of tube-like structures in a meteorite from Mars, possible remains of primitive bacteria-like organisms. See text for details of scale.

Nonetheless, the tube-like structures in Figures 4.13 *could* be the microscopic fossils of primitive bacteria-like martian organisms. They are claimed by the NASA scientists to be 'very similar in size and shape to very small microfossils found in some terrestrial rocks'. They are located in what are described as orange-coloured 'globules' of calcium, magnesium and iron carbonate within the meteorite. The structure and chemical composition of the 'globules' suggests that they formed by precipitation in an aqueous environment, possibly with the assistance of the primitive bacteria-like organisms which may be fossilized in the rock (Figure 4.13); i.e. the carbonates may be biogenic. However, the scientists have not yet been able to establish just how the carbonate 'globules' got into the rock.

The rock itself was found during an American National Science Foundation meteorite-collecting expedition in 1984. It is a portion of a meteorite that was dislodged from Mars by a huge impact about 16 million years ago and fell to Earth in Antarctica about 13 000 years ago, as determined by a variety of radioactive dating methods. These also give the age of the rock itself as 4.5 billion years, but place the age of the carbonate 'globules' within a rather large range of 3.6 to 1.4 billion years.

- Does the upper limit (older end) of this age range for the rock, the nature of the possible fossils, and the material in which they are found suggest any similarity with the remains of primitive life on Earth?
- If you refer back to Section 3.2.1, you will see that the combination of bacteria-like fossil structures associated with calcium carbonate, and an age of 3.6 Ga (i.e. close to 3.5 Ga), are all features of ancient stromatolites. Small bodies of shallow water ('warm little ponds') – or indeed even martian oceans, according to some scientists – may have provided an environment in which life on Mars could have developed.

The minute size of the structures in the martian meteorite must, however, give us pause for thought. So must the possibility that they are not in fact 3.6 Ga old, but could be as young as 1.4 Ga.

- Why would an age as young as 1.4 billion years give grounds for doubting the biological origin of the structures in Figure 4.13?
- By 1.4 billion years ago, more than two-thirds of the Earth's history had elapsed, life was abundant and widespread, and though probably still exclusively unicellular, there was a breathable atmosphere (see Section 4.7.1) and the eukaryotes (Box 4.3) were well established. Similarly (but sadly) Mars was also more than two-thirds of the way through its history, and must by then already have begun to resemble the planet we know today. The chances of even primitive life-forms developing under such conditions must be rated as small.

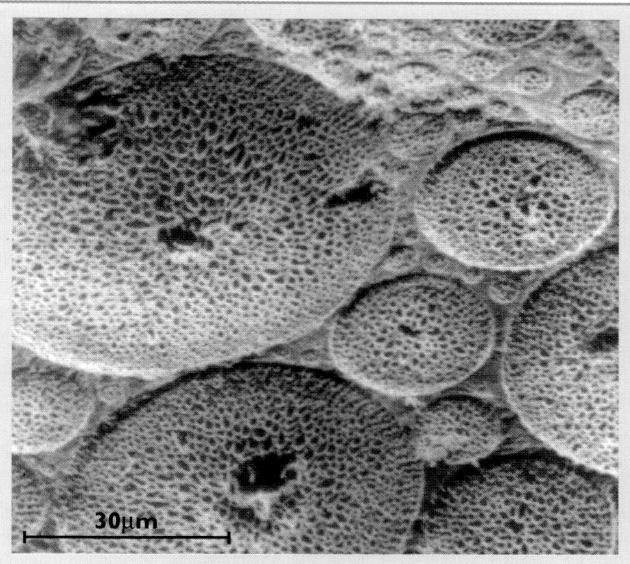

Figure 4.14
Intricately patterned honeycomb-like structures formed of aluminophosphate and synthesized inorganically in the laboratory. Note the difference in scale compared with Figure 4.13.

So might these fossils turn out to be of inorganic origin after all? The idea is not so far fetched, given the known properties of *amphiphiles* (Section 4.5). The structure in Figure 4.14 certainly looks as though it could have had a biological origin. It resembles the skeletal framework of silica (SiO_2) formed by some species of *diatoms*, which are aquatic algae (phytoplankton). In fact, it was made in the laboratory and is not composed of silica at all. It is an aluminophosphate compound formed simply by heating and recrystallizing an assemblage of simple organic and inorganic compounds, including an alkylamine amphiphile. Admittedly the structure in Figure 4.14 is huge in comparison with those in Figure 4.13, but while the actual mechanism is not fully understood, it seems to be what one authority in this field has called 'the natural extension of templated patterning of inorganic materials at the molecular and nanometre scales'. In other words, we cannot yet exclude the possibility that the 'martian microbes' may be nothing more than rather complicated microstructures of inorganic origin. We can, however, be confident that these martian meteorites have not been accidentally, still less deliberately, contaminated with terrestrial material – in other words, whether these 'fossils' prove to be the remains of ancient martian life or not, they are not a hoax (cf. Section 4.2.1 and Box 4.2).

More than that we cannot say at this point, but by the time you read this book, a great deal more will be known about these discoveries. Such carbon isotope data as were available in the mid-1990s gave inconclusive results. Who knows, 'white worms' may yet be found on Mars, though it is unlikely. That is because tube worms (Figure 3.17b) are animals – albeit very primitive ones – and for even primitive animals to have appeared on Mars, life there would need to have persisted for something like a couple of billion years or more, if the history of life on Earth is any guide. In fact, if the structures in Figure 4.13 turn out to be only 1.4 billion years old after all, the chances of any animal life ever having developed on Mars are very slim indeed.

4.7 The long haul begins

We will probably never know how many times life may have begun on Earth before it managed to gain a foothold – in other words, how many 'windows of opportunity' there really were. All we can say is that the first definite fossils date from 3.5 Ga ago and that sediments 3.8 Ga old *may* have been associated with living organisms (Section 3.2.2). Life could actually have become established even earlier, perhaps 4 billion years ago and perhaps more than once, only to be wiped out by a gigantic impact. Do we really *need* to know? Would it be sensible to conclude that life on Earth could only become established after the end of the major impact events 3.8 billion years ago and be content with that? Perhaps such questions are not so important, when set against the fact that fossils from a dozen or so well-documented sites in various parts of the world (and there are no doubt more such sites to be found) provide us with good evidence that life was widespread on Earth by the end of the **Archean** – the geological era from 3.8 to 2.5 billion years ago (Section 3.2.1).

It was all unicellular prokaryotic life, of course. Another couple of billion years or so elapsed before eukaryotic organisms made their appearance – the plants and animals familiar to us today are their descendants (Box 4.3). Prokaryotes comprise both producers (autotrophs) and consumers (heterotrophs) and while autotrophs probably dominated from about 3.5 Ga onwards, the metabolism of the very *first* organisms may have been heterotrophic (Section 3.2.1), an idea that originated with Oparin and Haldane (Section 4.2.2) and still has its adherents today.

■ How is such a scenario possible? By analogy with plants and animals, how could heterotrophs feed, i.e. from what could they obtain their energy for growth and replication, if there were no autotrophs to produce the food?

■ If you look back to the definition of heterotrophy in Section 3.2.1 you will see that it involves consumption of organic molecules, which can of course be produced abiotically, without the intervention of life.

There is abundant evidence that quite complex organic molecules must have been available on the early Earth, certainly supplied from comets and meteorites, possibly also synthesized by electric discharges in the atmosphere. It has been suggested that the very earliest organisms were methanogenic archebacteria (Box 4.3, cf. Figure 3.7) of the kind that still exist today, and that they would have been able to decompose simple organic molecules such as methanoic acid (HCOOH, formic acid) or ethanoic acid (acetic acid) to methane and carbon dioxide:

$$CH_3COOH \longrightarrow CH_4 + CO_2 \qquad \text{(Equation 4.3)}$$

This form of anaerobic decomposition is called *fermentation,* well known to anyone familiar with brewing or sewage treatment, and it occurs in all swamps and rice paddies, as well as in the digestive systems of many animals. Some modern archebacteria use a variant of this process to reduce carbon dioxide with hydrogen, which can be thought of as a form of autotrophic methane production (methanogenesis):

$$2CO_2 + 6H_2 \longrightarrow CH_2O + CH_4 + 3H_2O \qquad \text{(Equation 4.4)}$$

This is another possible metabolic mechanism for the earliest methanogens, but although the atmosphere was rich in carbon dioxide, it probably had little hydrogen (Section 4.4.1). Opportunities for the kinds of anaerobic metabolism summarized in Reactions 4.3 and 4.4 must therefore have been limited, especially as organic molecules were only available as a result of abiotic synthesis (including those supplied by comets and meteorites). Natural selection would tend to favour autotrophic systems that could supply their own reduced molecules for metabolism.

- What early autotrophic system has already been suggested?
- Photosynthesis based on the reduction of CO_2 via decomposition of hydrogen sulfide, H_2S, supplied by hydrothermal vents was discussed in Section 4.4.2 (Equation 4.2).

Radiation from the relatively weak late Hadean to early Archean Sun would have been capable of 'driving' these reactions, because it takes less energy to dissociate H_2S than it does for H_2O. The presence of sulfate in association with the oldest stromatolites (Section 3.6.2 item 5) is at least consistent with the proposition that the most ancient positively identified life-forms (Figure 3.6), 3.5 Ga old, were anoxygenic photosynthesizing organisms that used processes such as Reaction 4.1 to synthesize carbohydrates.

Supplies of organic molecules for heterotrophic methanogens would thus have been assured and methane could well have accumulated in the atmosphere. Methane is an even stronger greenhouse gas than is carbon dioxide and removal of CO_2 from the atmosphere (and hence of its greenhouse warming effect) to make organic molecules would have been offset by the build-up of atmospheric methane. If this had not taken place, surface temperatures might have fallen too far, because of the weak solar radiation that then prevailed.

However, autotrophs must also respire (which is what we do when we breathe). As you read in Box 1.5, carbohydrates are the fuel for metabolic processes. They are 'burned up' in respiration, liberating chemical energy that the organism can then use to synthesize other molecules such as the amino acids and more importantly the complex proteins without which growth, self-replication and reproduction are impossible. The other essential elements (nitrogen, phosphorus and sulfur, as well as others listed in Table 1.1), required to build these more complex molecules, are taken up by the (autotrophic) organism from water or air. However, the energy to build these molecules must come from the decomposition of the carbohydrate molecules that were themselves synthesized using sunlight, CO_2 and (in this case) H_2S. Where there is no oxygen, anoxygenic anaerobic autotrophs must 'respire' by fermentation, using reactions such as Equation 4.3 and the anaerobic decomposition of carbohydrate:

$$2(CH_2O)_n \longrightarrow nCH_4 + nCO_2 \qquad \text{(Equation 4.5)}$$

However, this is a much less efficient way of producing energy than is aerobic (oxygen-using) respiration, expressed at its simplest by the 'reverse' of the photosynthesis reaction (Equation 1.4):

$$(CH_2O)_n + nO_2 \longrightarrow nCO_2 + nH_2O \qquad \text{(Equation 4.6)}$$

But that is impossible in the absence of free oxygen. So for virtually all of the Archean (and possibly even longer) it seems that fermentation was so widespread and profligate that photosynthesis could do little more than keep up with it. Available evidence thus suggests that the earliest forms of life on Earth, including those that formed stromatolites, were photosynthesizing fermenters, to which atmospheric oxygen would have been toxic (as you probably know, the easiest way to stop any fermentation is to introduce air to the system, which kills the bacteria).

However, there are plenty of scientists who contend that the resemblance between the oldest fossils and modern cyanobacteria is not a matter of appearance only, but also of function – that these oldest Archean organisms (Figure 3.6) were oxygenic photosynthesizers, like their modern counterparts. But these organisms would not only have been incapable of aerobic respiration, they would have been poisoned by even quite small amounts of oxygen in their environment, just as fermenting bacteria are.

- So how might the oxygen produced by oxygenic photosynthesis be removed?
- Any free oxygen liberated into the Archean atmosphere would have been rapidly sequestered in oxidation reactions with exposed rock surfaces (especially the iron minerals and sulfides in those rocks), with reduced ions in solution in seawater (especially Fe^{2+}) and with atmospheric dust particles. So it would not have been available to poison organisms.

There is not yet a consensus on just when oxygenic photosynthesis began. We have just discussed one end of the spectrum of opinion, namely that it was in place 3.5 Ga ago – and some scientists would claim that the $\delta^{13}C$ record from the Isua rocks (Figure 3.7a) suggests even 3.8 Ga ago, but the data are consistent with anoxygenic photosynthesis too, cf. Figure 3.7b. Those who maintain that anoxygenic photosynthesis prevailed in the Archean provide differing estimates of when it changed to become oxygen-producing. These estimates range from 3 Ga, through 2.8 Ga to as little as around 2 Ga ago (well after the Archean), though it is probably fair to say that most would agree the change occurred some time *during* the Archean, i.e. before 2.5 Ga ago.

4.7.1 From fermentation to respiration

Once oxygenic photosynthesis had become the dominant autotrophic mechanism and life became more widespread on Earth, oxygen must have begun eventually to build up in the atmosphere. Some organisms then found a way of managing aerobic respiration, which is much more efficient than fermentation at using carbohydrate 'fuel'. It is difficult to overestimate the importance of cellular respiration on evolution. When fermentation and photosynthesis were coupled, organisms became merely self-sustaining – when photosynthesis became coupled with respiration the material of organisms could be used with such enormously greater efficiency that organisms were able to thrive. Life, having got a foothold, proceeded irrevocably to alter the surface of the Earth.

It seems unlikely, however, that aerobic respiration could really have become established until there was significant accumulation of oxygen in the atmosphere. Estimates vary, but most scientists would agree that before about 2 billion years ago the concentration of oxygen in the Earth's atmosphere was no more than a hundredth of its present level of 21% (i.e. it was less than 0.2%, of the Hadean and Archean atmosphere). Further discussion of these issues is outside the scope of this book, though it is important to stress that:

1. Oxygenic photosynthesis *may* have been the predominant autotrophic mechanism from 3.5 Ga, even 3.8 Ga ago, but it *could* have developed as late as the end of the Archean.
2. Anaerobic respiration (fermentation) almost certainly prevailed among both autotrophs and heterotrophs until around 2 Ga ago, when oxygen began to accumulate in the atmosphere. Only then was it feasible for aerobic respiration to become widespread.

It is perhaps worth noting that all eukaryotic organisms (Box 4.3) are what are known as obligate aerobes, that is to say they *require* oxygen for respiration and will die if deprived of it. According to some researchers, the reason why there could be no eukaryotes on Earth before about 2 billion years ago is that there was insufficient oxygen in the atmosphere to enable them to breathe. Only when oxygen levels began to rise, could the eukaryotes appear.

We must also address briefly the question of how oxygen could actually accumulate in the atmosphere, once aerobic respiration became a dominant feature of life on Earth. Reactions 1.4 and 4.6 can, in fact, be considered to be the 'reverse' of one another:

$$nCO_2 + nH_2O \underset{\text{respiration}}{\overset{\text{photosynthesis}}{\rightleftarrows}} (CH_2O)_n + nO_2 \qquad \text{(Equation 4.7)}$$

For every mole of oxygen liberated in photosynthesis, another mole is used up in respiration.

- So how could oxygen ever accumulate in the atmosphere?
- This could happen only by the *removal* of carbon in organic matter from the system, so that it could not be oxidized back to CO_2.

This process is only achieved in anoxic (anaerobic) environments, such as swamps, where decay of organic matter is never complete, leaving a legacy of carbon-enriched materials – which is why we have fossil fuels (coal, petroleum and natural gas). The oxygen that would have been used up in recycling that organic matter back into its constituents remains unused and so can accumulate in the atmosphere – though some, perhaps even most, of it will be removed in oxidation reactions associated with weathering of rocks and minerals at the Earth's surface.

4.7.2 The Earth System and its cycles

By the end of the Hadean, the Earth had a core, mantle and crust, a hydrosphere and an atmosphere, and by the end of the Archean there was a well established biosphere too; although that biosphere must have been largely (if not entirely) confined to the hydrosphere – it did not also permeate the atmosphere and upper parts of the crust. In other words, what is nowadays often called the **Earth System** was already functioning much as it does now. Most of the cycles we recognize today were well established, albeit in most cases on different time- and space-scales. Chief among these cycles was the water cycle of evaporation, precipitation, run-off and infiltration, which is intimately linked to the surface geological cycles of weathering, erosion and sediment deposition. Both of these are closely linked to the plate tectonic cycle (Figure 3.3) which continually recycles rock materials between the crust and the mantle. These major cycles are responsible for the physical movements of material between atmosphere, hydrosphere, crust and mantle. They in turn interact with, and are modified by, a whole suite of *biogeochemical cycles* that redistribute elements between different parts of the Earth System.

Chief among the biogeochemical cycles is the *carbon cycle* and there are analogous cycles involving all of the biologically important elements (nitrogen, phosphorus, etc.). The continued operation of these cycles unquestionably contributes to the maintenance of life on Earth, for as you read in the last sentence of Section 1.1.1 'although every individual creature must eventually lose its own battle with chaos, ... life itself continues.' The only reason life can continue is that when an organism dies its constituents at once become available for recycling into new organisms –

the energy needed for this renewal comes mostly from the Sun (and partly from the Earth's interior). These cycles are not closed, however, and their rates may change with time. For example, the plate tectonic cycle has certainly slowed with time because the interior of the Earth has cooled, as more of its radioactive heat-producing elements decayed away. Material has also been lost from the cycle, as the area of continental crust has grown from a negligible amount 3.8 billion years ago to cover about 30% of the Earth's surface now.

Another example of a cycle not being closed, is the progressive loss of carbon from the carbon cycle, which is why oxygen has accumulated in the atmosphere (Section 4.7.1).

- What two processes remove carbon from this cycle and what effect has carbon removal had on the surface environment of the Earth?
- The deposition of calcium carbonate as limestone – chiefly by marine organisms – has acted together with the removal of organic carbon into sediments, to remove CO_2 from the atmosphere. This has decreased the greenhouse effect that would otherwise have become stronger as the Sun has grown more powerful.

Time-scales are a crucially important aspect of the interaction between cycles. The time-scale of biological cycles for example, is years, and this is therefore the time-scale which governs the removal of a small proportion of organic carbon into the soil or sediments (i.e. out of contact with the atmosphere) when organisms die. For each mole of carbon (C) 'preserved' in this way, a mole of oxygen (O_2) remains free in the atmosphere. The time-scale of the geological and the plate tectonic cycles is such that the sediments receiving this carbon can continue to accumulate for thousands or even millions of years. Some of the oxygen freed by removal of this carbon remains in the atmosphere and some is used in the oxidation of rocks during weathering. Eventually, perhaps after tens or hundreds of millions of years, the sediments are returned to the Earth's surface and exposed once more to atmospheric weathering and oxidation.

The time-scales of these and other terrestrial cycles are of course dwarfed by those of the cosmic cycles involving the births and deaths of galaxies, stars and planetary systems, which are in the order of billions of years. It may not be too far-fetched to suggest that we owe our very existence to a small component of one part of one of these cosmic cycles. The principal reason why Earth is such a successful life-support system is its position in the Solar System, which has ensured that our planet has received just the right amount of solar radiation. Our neighbour on one side, Mars, is both smaller than the Earth and further away from the Sun. It receives much less energy from the Sun (its average surface temperature is –50 °C) and its weaker gravitational field meant it could not retain its original atmosphere. As you read in Section 4.6, recent research suggests that there was liquid water on Mars at one time and that this was 'carbonated' water (a consequence of the CO_2-rich atmosphere), from which calcium carbonate was precipitated (much like the scenario we sketched for the Hadean ocean on Earth). If life did emerge on Mars, it did not flourish and may not even have survived, perhaps because temperatures were too low and/or because the planet was too far from the Sun for solar radiation to be intense enough to sustain it.

Our other neighbour, Venus, which is closer to the Sun and about the same size as the Earth, has retained its CO_2-rich atmosphere but lost all its water – its average surface temperature is around 460 °C. Could Venus once have had an ocean, in which life might have emerged? If so, it did not last long enough to remove CO_2 from the atmosphere and so moderate the increased greenhouse warming that would have occurred as the Sun became stronger.

- To a first approximation could the Earth be considered to have become a closed system (i.e. experiencing neither gains from nor loses to its surrounding space) at the end of the Hadean?
- Only in the sense that by about 3.5 Ga ago, perhaps even 3.8 Ga ago, the overall bulk composition of the Earth was much as it is today, i.e. it had received virtually its full complement of chemical elements.

As we have already mentioned, the Earth System continues to be supplied with extraterrestrial debris from meteorites and comets, but this flux became greatly reduced after about 3.8 billion years ago and is presently minute relative to the mass of the Earth. There have been some losses too, in the form of hydrogen and other light gases escaping from the Earth's gravitational field, but these are also small in the context of the Earth as a whole. To a good first approximation, the Earth's total complement of chemical elements has not changed greatly since the end of the Hadean.

In terms of energy, however, the Earth is not a closed system at all. A great deal of heat comes from the Earth's interior and without this there would be no plate tectonic cycling between the crust and the mantle. Without the Sun there would be no water cycle and no surface geological processes of any significance – above all, there would be no life. We must not forget gravitational effects either – interactions with other bodies in the Solar System, especially the Sun and the Moon, control the tides and have progressively slowed the Earth's orbit and rotation rate through geological time.

It is essential to bear in mind that in the Earth System (indeed in any system) 'everything relates to everything else'. Change in one part of the system can affect other parts, in often unpredictable ways. Interactions that are not always anticipated (and often not even suspected) include those related to human activities. For example, huge quantities of sulfur dioxide are annually released into the atmosphere by industrial processes, especially in the Northern Hemisphere and it is widely believed that this leads to formation of sulfate aerosols and acid rain. However, it is not so well known that those same sulfate aerosols may scatter sunlight so effectively as to offset the effects of global warming caused by increased concentrations of the greenhouse gas, CO_2, also from industrial activities. Another example is the concern of some scientists that continued global warming could lead to changes in circulation of the North Atlantic. This could prevent the Gulf Stream from reaching northwest Europe (including Britain), with the result that feedbacks (Box 3.2) associated with global warming could have the paradoxical effect of actually causing annual mean temperatures to *fall* by several degrees in a region accustomed to mild winters and warm summers.

- In what way might the present-day surface environment of Venus, described above, be considered to represent a sort of apocalyptic warning to humanity if it continues its profligate use of fossil fuels?
- If atmospheric CO_2 levels continue to rise exponentially at rates that appear to be far greater than at any time in the whole of Earth's history, the enhanced greenhouse effect could lead to runaway global warming, exacerbated by a variety of feedback mechanisms, one of which would be melting of the polar ice-caps.

The intimately interlocking relationships of the myriad components of the Earth System began to be generally recognized only when space travel enabled everyone to see a picture of the Earth 'floating' in space (Figure 1.1). The enormous changes that humans have made to the face of the planet have been for the benefit of one part of the Earth System only – humanity itself. One of the greatest of these changes has been the manipulation of the blueprint of life through genetic engineering during the 1990s. Proliferation of environmental movements and groups in recent years shows that we are now more aware of the dangers of ignoring parts of the global system which may not appear to be of immediate concern to us.

Fascinating though such issues are, however, they remain outside the scope of this book. As the saying goes – all that is another story.

4.8 Summary of Chapter 4

1 The essential elements of life, carbon, hydrogen, oxygen, nitrogen and phosphorus, are among the most abundant in the cosmos. The complex molecules necessary to construct self-replicating life-forms develop under only a relatively narrow range of conditions, the chief requirement being the presence of liquid water (i.e. at temperatures greater than 0 °C and less than 100 °C).

2 Theories about the origin of life have included spontaneous generation and panspermia, both now discarded in favour of some kind of chemical evolution. Life is thought to have arisen abiotically (abiogenically) from organic molecules which may have been synthesized on Earth and/or in space, in the latter case being brought to Earth by meteorites and comets.

3 The first organisms to evolve on Earth were prokaryotes (bacteria) which are unicellular. Eukaryotes evolved much later. They include all plants and animals but some are also unicellular. Organisms are based on cellular units and the properties of genetic information storage and replication reside in the nucleic acid, DNA. The information store (genetic code) consists of four nucleotide bases: the purines adenine and guanine (A and G), and the pyrimidines cytosine and thymine (C and T), which are linked by sugar and phosphate groups in the double-helix structure of DNA. The bases in the two strands of the DNA double helix are linked according to the A–T G–C pairing rules.

4 Laboratory experiments have synthesized complex organic molecules by passing electric discharges through a variety of starting materials intended to represent the early ocean and atmosphere. The largest yields of amino acids and related compounds were obtained when reduced gases were used (methane, ammonia, etc.).

5 Ideas about the origin of life on Earth include the proposal that hydrothermal vent environments could have provided a suitable setting. In the absence of oxygen, chemosynthesis of organic molecules would have been achieved via the reduction of carbon dioxide using hydrogen from the dissociation of hydrogen sulfide supplied in hydrothermal vent solutions with solar radiation as the energy source.

6 The variety of organic compounds, including amino acids, found in carbonaceous chondrites suggests that the building blocks of life may have been supplied to Earth from space. The chirality of organic compounds in meteorites (including amino acids) points unequivocally to an origin in space

and excludes contamination within the Earth System. It is possible that life may have originated on other planets, especially Mars, early in the history of the Solar System.

7 Amino acids and other compounds would need to have become sufficiently concentrated in order to be able to combine into complex molecules (nucleic acids) capable of self-replication. Possible substrates upon which this might have happened include the surfaces of clay minerals or of sulfide particles.

8 The question of whether DNA or RNA was the first to appear in primitive organisms remains unresolved. Some types of RNA can replicate themselves, i.e. they can provide their own 'blueprint'. In the earliest Archean, primitive life may have inhabited an 'RNA world' which was soon supplanted by the 'DNA world' which has dominated all life-forms since.

9 In the oxygen-free environments of the Hadean Earth, metabolism of the earliest organisms must have been centred round (anaerobic) fermentation rather than (aerobic) respiration. The earliest autotrophs probably resembled anoxygenic photosynthesizing methanogens but the possibility of oxygenic photosynthesis cannot be ruled out. Either way, free oxygen would have been toxic to all organisms. It is not known how or when the transition to oxygenic photosynthesis occurred, but only then was it possible for atmospheric oxygen to begin to accumulate, by removal of small amounts of organic matter from the carbon cycle. Aerobic respiration then became possible, but atmospheric oxygen concentrations had to reach significant levels before eukaryotes could evolve.

10 The Earth System is not a closed one: the Sun provides most of the Earth's energy, gravitational and electromagnetic energy are exchanged with other parts of the Solar System, and the Earth continues to receive cometary and meteoritic material from space.

Now try the following questions to consolidate your understanding of this chapter.

Question 4.7

What are the principal differences between 'spontaneous generation' and the theory of chemical evolution of life?

Question 4.8

Which of the following statements about the early Earth are true?

(a) The Hadean Era was from 4.6 to 3.8 million years ago.

(b) For most of the Hadean, the Earth's atmosphere contained significant amounts of ammonia (NH_3) methane (CH_4) and hydrogen (H_2).

(c) There was no ozone layer in the Hadean atmosphere.

(d) Rocks from the Hadean provide evidence that the Earth's surface was much warmer in the Hadean than it is now.

(e) The Hadean atmosphere probably contained up to 5% free oxygen.

Question 4.9

Would you say that the atmosphere of Titan (Section 4.4.3) as deduced from information provided by the Voyager spacecraft, resembles the Earth's 'pre-Moon' atmosphere, (i.e. the one that existed in the earliest Hadean soon after formation of the Earth)? Which of the laboratory experiments described in Section 4.4.1 seem to have replicated these conditions most closely, those of Miller and Urey, or those of later workers?

Question 4.10
What effect might the continued large-scale combustion of fossil fuels have on the $\delta^{13}C$ of plants that photosynthetically fix CO_2 from the atmosphere?

Question 4.11
In a sentence or two, explain (a) why land areas in the Hadean might have looked something like Figure 4.15; (b) how Figure 4.16 illustrates one kind of environment in which life could have originated; and (c) what the life-forms might have been in such an environment.

Figure 4.15
Volcanic island terrain. Basaltic lavas (black) occupy part of the middle ground and form most of the headland behind, the rest is formed of volcanic ashes (brown, layered).

Figure 4.16
Hot springs in volcanic terrain. The rocks are coated with deposits of silica (grey area), a little sulfur (yellow area) and probably with compounds of iron (brown area).

Objectives

When you have finished this book, you should be able to display your understanding of the terms printed in **bold** type within the text as well as the following topics, concepts and principles and, where appropriate, perform simple calculations related to them:

1. the disequilibrium between all life and its surroundings and the dominance of carbon, hydrogen, nitrogen and oxygen in all known life-forms;
2. the central role of carbon in forming the complex self-replicating molecules essential to life;
3. the distinction between a nuclear fusion reaction and a chemical reaction;
4. the significance of the stellar CNO cycle for 'life-forming' elements;
5. the source regions of asteroids, meteorites and comets;
6. the cosmic abundance of the elements and the basis for estimating them, including the use of carbonaceous chondrites;
7. the contrasts between rocky and gaseous planets in relation to the condensation sequence of elements and compounds in the solar nebula;
8. the evidence from the oldest rocks, stromatolites, carbon isotopes, lunar history, and other (circumstantial) evidence for conditions on (and within) the early Earth;
9. the observational and experimental basis for hypotheses about the origin of life on Earth;
10. the A–T G–C pairing rules of the genetic code;
11. the transition from anoxygenic photosynthesis (or chemosynthesis) and anaerobic fermentation to oxygenic photosynthesis and aerobic respiration;
12. the Earth as a system of interlocking cycles in which 'everything relates to everything else'.

Answers to Questions

Question 1.1

Crystals cannot replicate themselves. Aggregates of crystals can form from solution, but once formed, each crystal remains a unique entity – the only way to get more crystals is to make more solution, and to start again.

Question 1.2

The silicon-based polymers all involve only Si–O bonds and form silicate minerals, whereas carbon forms bonds not only with itself but also with hydrogen, oxygen, nitrogen and other elements and produces molecules that are far larger and more complex than any silicate. What is more, we know from our own experience on Earth that self-replicating molecules are carbon-based and not silicon-based.

Question 2.1

We start by rearranging the equation $E = mc^2$ so that it is in the form:

$$m = \frac{E}{c^2}$$

We then substitute the values given as follows

$$m = \frac{10^{-12}\,\mathrm{J}}{9\times10^{16}\,\mathrm{m^2\,s^{-2}}} = \frac{10^{-12}\,\mathrm{kg\,m^2\,s^{-2}}}{9\times10^{16}\,\mathrm{m^2\,s^{-2}}} \approx 10^{-29}\,\mathrm{kg}$$

This is the amount of mass converted into energy per individual H–D fusion.

Question 2.2

They would have been identical. Deuterium is an isotope of hydrogen and so it behaves in an identical way chemically. Thus the deuterium would have combined explosively with oxygen to produce water; although the water would have been ^2H_2O (D_2O), otherwise known as 'heavy water'.

Question 2.3

The elements are not found naturally because all nuclei heavier than those of uranium are unstable and rapidly disintegrate by radioactive decay to form nuclei (and hence elements) of smaller mass.

Question 2.4

(a) The chemical elements are synthesized in stars. In general, elements with masses less than that of iron are formed by nuclear fusion reactions. Simple examples include the fusion of hydrogen nuclei to give helium, or the fusion of helium nuclei to form carbon.

(b) Elements heavier than iron are formed by neutron capture operating either on a slow time-scale during the star's lifetime, or rapidly in a supernova explosion.

(c) Carbon is principally synthesized in stars by the fusion of three helium nuclei to form ^{12}C (carbon-12).

Question 2.5

No. Hydrogen and helium dominate the composition of the Sun, but would not be expected to be abundant in a solid object like a meteorite – even in a carbonaceous chondrite; although some hydrogen must be present there, combined with carbon, nitrogen and oxygen in organic molecules (and as H_2O in water). Hydrogen and helium would plot so far above the line in Figure 2.11 as to be off the scale of the graph.

Question 2.6

(a) The carbon in the Sun was formed by nuclear fusion in other stars. It was not formed within the Sun itself, so the Sun acquired it 'secondhand'.

(b) Carbon is expelled (along with other elements) from the Sun in the solar wind, and some of it has reached the Moon, where its products have been sampled by astronauts on lunar missions. The solar wind does not reach the Earth's surface because of the strong magnetic field.

Question 2.7

Water in comets was originally formed by ion–molecule reactions in molecular clouds (Section 2.4.2, Reaction 2.4). The water molecules condensed as ice onto the surfaces of interstellar dust grains, which in turn aggregated into larger icy bodies.

Question 2.8

The star is presumably in the early stages of the CNO cycle (Figure 2.5): ^{12}C (carbon-12) nuclei would fuse with hydrogen nuclei (^{1}H, hydrogen-1 in this case) to form the larger (heavier) nuclei of nitrogen and oxygen, liberating helium nuclei in the process, i.e. hydrogen is being converted into ^{4}He (helium-4) not only by hydrogen fusion in the body of the star but also through the catalytic action of the CNO cycle.

Question 2.9

(a) A star formed early in the history of the Universe would be composed only of hydrogen plus some helium. We have seen that neither of these gases figure prominently in planetary formation (they are in essence too volatile, i.e. their melting/freezing temperatures are too low), so we would not expect planets to form.

(b) Even if such planets did form, they would be incapable of supporting life because that requires complex molecules, which can be provided only by a range of heavier elements, especially carbon.

Question 2.10

(a) Mars is the same age as the Solar System, about 4.6×10^9 (4.6 billion) years old.

(b) Mars is a solid (rocky) planet, one of the four inner planets, whereas Jupiter is essentially a large ball of gas.

(c) No. Mars contains a high proportion of the refractory (higher melting) elements and compounds because it formed nearer to the Sun. It has fewer of the volatile (gaseous) constituents, which are more abundant on Jupiter, which formed further from the Sun.

Question 2.11

The elements are synthesized in stars, by nuclear fusion (oxygen is formed in the CNO cycle) and neutron capture (for heavier elements). In contrast, the formation of aluminium oxide involves a chemical reaction, of the kind that we are familiar with – except that it takes place in the near-vacuum of space, albeit in the outer envelopes of stars where temperatures are sufficiently high for such reactions to occur.

Question 2.12

(a) Asteroids are mainly fragments of rock, orbiting near the inner planets (between Mars and Jupiter, Figure 2.7). They formed at the same time as the rest of the Solar System (4.6×10^9 years ago). Comets come from the Oort cloud and the Kuiper belt, in the outer reaches of the Solar System. They are formed of dust particles/grains, condensed gases and ice, together with a variety of complex organic molecules. They may be remnants of the dense molecular cloud that was the forerunner of the Solar System.

(b) The solar nebula was hottest at its centre. The rocky planets accreted from the more refractory higher-temperature materials in the central region, from which heating had driven off most of the volatile (gaseous) components. The outer planets, formed in the cooler regions of the nebula, would have retained a much higher proportion of their accreted volatile matter.

Question 2.13

The maximum and minimum figures for the age of the Universe are, respectively, 20×10^9 and 10×10^9 years.

Question 3.1

Any free oxygen in the atmosphere would be taken up in reactions with rocks and minerals at the Earth's surface, unless there were a source of supply that produced oxygen faster than it was being used up. The only source of such supply that we know of is life. See the text following Question 3.1 for further discussion.

Question 3.2

A $\delta^{13}C$ of –50‰ is outside the limits for any of the organisms in Figure 3.7b. See the text following Question 3.2 for further discussion.

Question 3.3

You could infer that because comets are 'dirty snowballs' rather than 'rocks' they would be more easily vaporized in the atmosphere. However, the impacts of the comet Shoemaker–Levy on Jupiter suggest this is not the principal reason. It is probably mainly because both the Kuiper belt and the Oort cloud (the sources of comets) are further from Earth than the asteroid belt.

Question 3.4

Volume of asteroid $= (4/3) \times 3.14 \times (250\ \text{m})^3 = 6.5 \times 10^7\ \text{m}^3$ (rounded down)

Mass $= 3000\ \text{kg m}^{-3} \times 6.5 \times 10^7\ \text{m}^3 = 1.95 \times 10^{11}\ \text{kg}$ (about 200 million tonnes)

$$\text{Kinetic energy} = 0.5 \times 1.95 \times 10^{11}\ \text{kg} \times (20 \times 10^3)^2\ \text{m}^2\ \text{s}^{-2}$$

$$= (0.5 \times 1.95 \times 10^{11} \times 4 \times 10^8)\ \text{kg m}^2\ \text{s}^{-2}$$

$$= 3.9 \times 10^{19}\ \text{J}$$

Question 3.5

The histogram in Figure 3.10b suggests at first sight that the incidence of collisions with extraterrestrial bodies has *increased* with time: there are nearly 50 known craters less than 50 Ma old, compared with a mere handful more than 600 Ma old. The reason is that older craters have been progressively weathered and eroded by geological processes, so very few ancient craters now survive.

Question 3.6

Yes. Even the earliest 'fruit-cake mix' Earth must have consisted mainly of solid refractory materials, which became rearranged or differentiated into the layered structure we see illustrated in Figure 3.2. There must have been some loss of volatile material to space along the way, but this is not likely to have made a significant difference to the overall bulk composition.

Question 3.7

(a) Most of the iron is in the core, probably alloyed with nickel (Figure 3.2).

(b) The lack of iron in the Moon cannot be due to an early heating event because iron is one of the heavier elements, and is most unlikely to be volatile enough to be expelled by heating – if it were, there would be no silicates or aluminosilicates either.

Question 3.8

Because the Highland rocks are well over 4 billion years old, they cannot have been geologically 're-processed' since that time.

Question 3.9

(a) The Highland area has a far greater number of craters per unit area than has the mare. In fact, virtually everywhere in the Highlands is either part of a crater or has small craters on it. On the mare surface, a high proportion of the area shows no resolvable craters, and the rilles have not been obliterated by craters.

(b) There are larger craters on the Highland surface, some more than 50 km across, whereas the mare surface has craters only up to about 20 km across.

Question 3.10

No, the record of the lunar craters shows that meteorites continued to arrive in near-Earth space in considerable numbers more recently than 3.8 billion years ago, and Figure 3.10 shows that they haven't stopped doing so yet. The big difference is that the really big impacts producing craters up to 5000 km in diameter appear to have ceased some 3.8 billion years ago.

Question 3.11

The density of basaltic rocks is great enough for them to be subducted back into the mantle (Figure 3.3 and related text).

Question 3.12

3.3 billion years ago is more than a billion years (>1000 Ma) after the proposed origin of the Moon and that age is 1.2 billion years younger than the oldest lunar rocks. Preservation of tidal structures in such sandstones indicates only that the Earth was subjected to tidal influences at that time, and you know from Section 3.6.2 (item 6) that the Sun also raises tides on Earth. Herringbone patterns provide no positive information about spring–neap tidal cycles, which would be necessary to provide evidence of lunar influence and can therefore give no support to any hypothesis about the origin of the Moon.

Question 3.13

If the radiation was 25% less it would have been 75% of its present value, i.e. $0.75 \times 1.5 \times 10^{17}$ W = 1.1×10^{17} W and radioactive heat production would have been $5 \times 1.7 \times 10^{13}$ W = 8.5×10^{13} W, still about three orders of magnitude less. Some extra heat must have been supplied by large impacts, and a good deal more from the early levels of primordial heat relatively soon after accretion and core formation (Section 3.1.2).

Question 3.14

The value for iron (Fe) on the vertical axis (the Sun) in Figure 2.11 is about 1.5×10^6. 25% of that is about 0.4×10^6, i.e. 4×10^5. The point would, in fact, be shifted only two or three millimetres to the left, which still leaves it closer to the line than the point for N (nitrogen) in carbonaceous chondrites. This is an instructive illustration of how graphs with logarithmic scales can accommodate considerable variations and still demonstrate good correlation between variables.

Question 4.1

There was no oxygen in the Hadean atmosphere and nearly all modern life-forms need oxygen for respiration. Life would indeed quite literally be 'snuffed out'. Without oxygen there could be no ozone to provide a shield against the harmful effects of ultraviolet radiation from the Sun – even though the Sun was 25–30% less intense than it is now and some radiation would have been reflected by clouds. Levels of ultraviolet radiation could well have been high enough to destroy any burgeoning life-forms not shielded by an adequate thickness of water (Section 3.6.1).

Question 4.2

Along with helium (He), which is an inert gas (and is therefore ill-equipped for a role in life processes), carbon, hydrogen, oxygen and nitrogen are the most abundant elements in the Universe.

Question 4.3

The densest regions of interstellar space would seem to be the giant molecular clouds, millions of billions of kilometres across, though even these contain only 10^9 to 10^{12} atomic or molecular particles per cubic metre.

Question 4.4

(a) We are told that about 1% of solar ultraviolet radiation has a wavelength of less than 200 nm and that solar ultraviolet radiation is about 20% of total solar radiation. So the power available for pre-biotic chemistry is calculated to be:

20% of 10^{17} W is $20 \times 10^{-2} \times 10^{17}$ W = 20×10^{15} W

1% of this is $1 \times 10^{-2} \times 20 \times 10^{15}$ W = 2×10^{14} W

(b) Ultraviolet wavelengths are now absorbed by the stratospheric ozone layer (Section 3.6.1). This ozone layer could only form after oxygen had accumulated in the atmosphere, which did not even begin to happen until at least a few hundred million years after the Hadean.

Question 4.5

(a) The greatest yield of amino acids and other complex organic compounds was produced in the Miller–Urey experiments, which used the most reduced constituents for the laboratory 'atmosphere'. All the other experiments, performed in more oxidized 'atmospheres', yielded only simple molecules of the kind that can be found in interstellar space anyway (Tables 2.1 and 2.2), which has a bearing on part (b) of this question.

(b) The implication of part (a) is that formation of complex organic molecules using stellar (solar) radiation as the energy source requires reduced compounds as starting materials. So, if the Hadean atmosphere was dominated by less reduced compounds, then abiotic synthesis could produce only simple organic compounds. Conditions on Earth may not have been suitable for turning these basic materials into the more complex molecules necessary for life to emerge and develop. We expand on this theme in the next section.

Question 4.6

3×10^5 t of cosmic dust falls to Earth each year. 5% of this (i.e. $3 \times 10^5 \times 5 \times 10^{-2}$ t $= 1.5 \times 10^4$ t) comes from carbonaceous chondrites.

Of this, 3% is made up of organic molecules. This is equal to $1.5 \times 10^4 \times 3 \times 10^{-2}$ t $= 4.5 \times 10^2$ t = 450 t organic matter per year to the Earth as a whole, which is 4.5×10^5 kg.

This falls over an area of $5 \times 10^{14}\,\text{m}^2$, so the average annual flux is $(4.5 \times 10^5\,\text{kg})/(5 \times 10^{14}\,\text{m}^2) = 9 \times 10^{-10}\,\text{kg m}^{-2}\,\text{yr}^{-1}$.

Question 4.7

Proponents of spontaneous generation (a widely held belief in the early days of scientific enquiry, in the 16th to 18th centuries, before rigorous observation and experiment were brought to bear on such questions) held that life literally arose spontaneously from nothing, e.g. worms in rotting apples, maggots in rotting meat. The theory of chemical evolution is that life was abiotically synthesized from simple molecules such as amino acids, which themselves had formed by abiotic reactions on the early Earth and/or in space (e.g. interstellar clouds) brought to Earth by meteorites and comets.

Question 4.8

(a) False. The word should be *billion* not million.

(b) False. The Earth's *original* atmosphere may have contained these reducing (and reduced) molecules, but it was probably lost when the Moon was formed. The subsequent Hadean atmosphere was probably dominated by nitrogen and carbon dioxide.

(c) True. Without free oxygen in the atmosphere, there could be no ozone layer; which makes (e) a false statement as well.

(d) False. There are hardly any rocks preserved from the Hadean and those that do exist provide no climatic information. However, it is considered likely that the Hadean Earth *was* probably a warmer environment than now.

(e) False. The Hadean atmosphere was devoid of oxygen (see (c) above).

Question 4.9

Information in Section 4.4.3 suggests that the atmosphere of Titan is a reducing one, with nitrogen combined in ammonia and carbon in methane. This is much more like what the Earth's pre-Moon atmosphere is inferred to have been; the post-Moon atmosphere is believed to have been dominated by nitrogen and carbon dioxide. The Miller–Urey experiments would have more closely replicated conditions on Titan, as they used reduced gases. Later researchers used mildly oxidizing/oxidized mixtures.

Question 4.10

The $\delta^{13}C$ of plants is *less* (more negative) than that of non-organic carbon, including atmospheric CO_2 because plants are relatively enriched in ^{12}C. Large-scale fossil fuel combustion will put relatively more ^{12}C than ^{13}C back into the atmosphere, theoretically making relatively more ^{12}C than ^{13}C available for take-up by plants. However, the difference is not likely to be great, possibly 1 part per thousand (1‰), which might not be detectable, given the ranges shown in Figure 3.7.

Question 4.11

(a) Land areas in the Hadean were mostly volcanic islands formed of basaltic lavas and probably of volcanic ashes also. They were devoid of any vegetation. As the Sun was weaker, the sky and sea may not have been so blue and there might have been more clouds.

(b) Hot springs on volcanic islands or near-surface hydrothermal vents are presently favoured as the likely environments for the origin of life on Earth (Section 4.4.2). Silica deposits would have been leached from the volcanic rocks by the hot waters, sulfur would have come from dissociated H_2S, see part (c), and the iron compounds could have included carbonate.

(c) Life-forms would have been anoxygenic photosynthesizing bacteria, breaking down hydrogen sulfide in the water to reduce CO_2 to form organic molecules, using solar radiation as the energy source.

The photographs in Figures 4.15 and 4.16 were taken on the Azores.

Comments on Activities

Activity 3.1

Your version of the caption to Figure 3.15 is unlikely to match the one below in detail (because there are likely to be as many versions as there are people doing this Activity), but you should see if you have captured the main points.

Figure 3.15

Schematic representation of accretion of planet Earth (or of an Earth-like planet) from planetesimals, asteroids, etc., its continued bombardment and eventual differentiation into core, mantle and crust.

(a) Small cosmic bodies and particles of cosmic dust accrete to form larger planetesimals and asteroid-sized bodies (b), which themselves accrete into a larger body (c), which is still very much smaller than the Earth. This in turn continues to grow by continued accretion of meteorites, asteroids and comets, to form (d) the early 'fruit-cake mix' Earth (Section 3.3), which also continues to be bombarded by meteorites and is expelling volatile material (outgassing, see Section 3.6.1). (e) Soon after the Earth reaches its present size, and during continued bombardment from outer space, the 'fruit-cake mix' Earth differentiates into a core, mantle and (possibly) crust, with the sinking of metal (mainly iron, with probably some nickel) to the centre, displacing lighter silicate minerals to form the mantle (mainly iron and magnesium silicates, see Figure 3.2) and crust (probably mainly basaltic at this time, as not much granite has yet formed at this early stage in the Earth's history). The diagram in (e) is a schematic cross-section depicting this process approaching completion, with the core and mantle being formed.

If you are familiar with the nature of minerals, you may feel the term 'light silicates' to be a misnomer in this context, as the silicates would be mainly iron and magnesium-rich (olivine and pyroxene) and relatively dense compared with the aluminium-rich silicates (feldspars) and quartz that dominate the crust. Some silicates are thus lighter than others, though all are lighter (less dense) than iron. But this is by the way and not central to our theme.

These events would all have taken place within about 0.1 billion years (100 Ma) of the formation of the Solar System, i.e. between about 4.6 and 4.5 billion years ago.

If (d) and (e) represent the approximate present size of the Earth, then a scale bar similar in size to that used in (c) would have the label 5000 km in these two parts of Figure 3.15, because the Earth's radius is over 6000 km.

Activity 3.2

The answers to the questions raised in Activity 3.2 are as follows:

1 The numbers on the scale span seven orders of magnitude. If you tried to plot that sort of range on linear graph paper, using, say, 1 mm for the interval between 1 and 2, your histograms would be 100 000 mm high, which is 100 m. If you used, say, 1 mm for the interval between 100 and 200, your histograms would still be 1000 mm high, which is 1 m; and all of the smaller intervals, below 100, would be crammed into the lowest millimetre of the scale.

Logarithmic scales are obviously useful for plotting large ranges of numbers.

2(a) The figure for hydrogen is close to 1×10^5 (i.e. 100 000), because it is the most abundant element in the Universe, and we can use that for the purposes of this question, (in fact the number is about 9×10^4, which makes little difference when you deal in orders of magnitude). For oxygen the number is a bit under 1×10^2 (in fact it is about 60, but again that is not a big deal in relation to orders of magnitude). So, in atomic terms there is about $(1 \times 10^5)/(1 \times 10^2) = 1 \times 10^3$ or 1000 times more hydrogen than oxygen in the Universe. If you want to be pedantic, the sum would be $(9 \times 10^4)/60$, or 1500 times, which is still 1.5×10^3, and is the same order of magnitude.

(b) The relative atomic mass of hydrogen is about 1, and that for oxygen is 16. In mass terms, then, we have $(1 \times 10^5 \times 1)/(1 \times 10^2 \times 16)$, which is $(1 \times 10^5)/(1.6 \times 10^3) = 0.6 \times 10^2$, or about 60 times more hydrogen than oxygen.

Even by mass, therefore, hydrogen is still more plentiful than oxygen in the Universe by nearly an order of magnitude.

3(a) Points that occur on the line represent elements that have the same relative abundance in the Universe and either the human body or the Earth's crust.

(b) The value for hydrogen (H) in the human body and the collective value for 'all other' elements in the human body fall closest to the line.

(c) It is the nature of logarithmic scales that they can never go to zero, but our scale does not go below 0.1, and the elements in Figure 3.24 with relative abundances below 0.1 are not shown on the histogram. We cannot know by how much they are less than 0.1 in relative abundance, hence the arrow indicating 'down there (or over there) somewhere'.

(d) Hydrogen is only slightly more abundant (relatively) in the Universe than in the human body, but it is relatively much less abundant in the Earth's crust than in either of the other two. (Remember these are *atomic* proportions – in terms of mass, hydrogen has low abundance in the crust cf. part 2b above.)

Aluminium is relatively hugely abundant in the Earth's crust relative to the Universe, and is not even on the scale so far as the human body is concerned. This should occasion no surprise, as much of the Earth's crust consists of aluminium-rich silicate minerals, while the human body is inclined to treat aluminium compounds as toxic substances.

Helium is far more abundant in the Universe than in either the crust or the human body. That is not surprising, for helium is almost completely unreactive chemically, and is most abundant in stars.

Silicon, somewhat surprisingly, is only slightly more abundant (in relative terms) in the Universe than in the human body, but is hugely more plentiful in the crust, where it is bound up in silicate minerals. Only some of these minerals contain aluminium, so silicon is the more abundant of the two.

4 See Figure A1 (this is the completed graph of Figure 3.25).

(a) It is not a totally convincing case. The 'Earth's crust' point for the major element of life, carbon, for example, is actually closer to the line than the 'human body' point for this element. For oxygen, the difference is small, but the 'crustal' value for nitrogen lies closer to the line than the 'human body' value. As we have seen, hydrogen plots almost on the line for the human body, well off it for the crust. Sulfur is also biologically important, but its 'crustal' value also lies closer to the line than its 'human body' value, as does that for phosphorus, which is also an important element for life.

The 'human body' values for magnesium, sodium, calcium, potassium actually lie closer to the line than their 'crustal' values, which does lend a little support to the statement given in the question. Interestingly, chlorine is actually relatively enriched in humans compared with the crust, which is perhaps not surprising as sodium chloride is a fairly prominent component of body fluids (blood, sweat and tears). Aluminium (see part 3d above) and fluorine are prominent crustal constituents that are not particularly life-friendly, and do not appear on the 'human body' histogram.

On balance then, it is by no means obvious that the cosmic abundance of the elements is more nearly represented in the human body than in the Earth's crust.

(b) Probably not. The elements C, N, O, H are essential for life, sulfur and phosphorus provide essential nutrients, and few organisms can get along without sodium, potassium, magnesium and calcium (refer to Table 1.1). So we might conclude that the histogram would not look all that different.

(c) Yes. The horizontal axes in Figure 3.24 show atomic numbers of the elements, and it is plain that the heights of the histogram bars decline with increasing number in part (a) of the figure. The relationship is less obvious for the other parts of Figure 3.24. That is because the rock and plate tectonic cycles on the one hand, and life processes on the other, have acted to fractionate the elements, and change their relative abundances away from the cosmic 'baseline'.

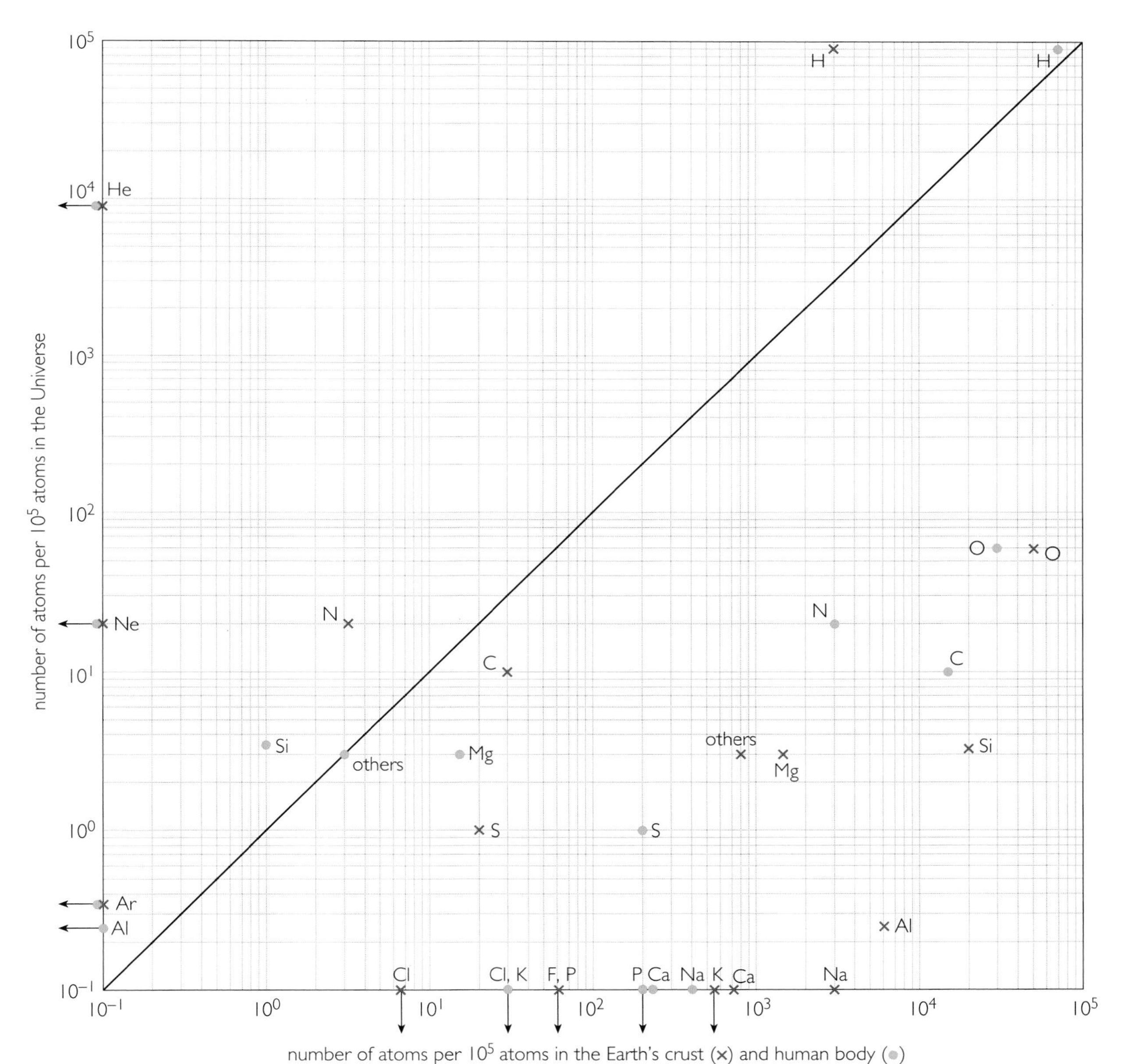

Figure A1 Completed graph for Activity 3.2. The data points you should have plotted are shown as crosses.

Appendix 1 The Periodic Table

The Periodic Table displays the elements in order of increasing atomic number. The large numbers (1–8) above the columns are the Groups, e.g. carbon is in Group 4. The individual element boxes contain the atomic number at the top, the relative atomic mass at the bottom, and the chemical symbol in the middle.

Elements with atomic numbers up to 112 have now been reported. However, the names of the elements with atomic numbers 104 to 112 (and therefore their chemical symbols) have not been settled, so the boxes have been left blank in this version of the table.

1	2											3	4	5	6	7	8
1 H hydrogen 1.01																	2 He helium 4.00
3 Li lithium 6.94	4 Be beryllium 9.01											5 B boron 10.8	6 C carbon 12.0	7 N nitrogen 14.0	8 O oxygen 16.0	9 F fluorine 19.0	10 Ne neon 20.2
11 Na sodium 23.0	12 Mg magnesium 24.3											13 Al aluminium 27.0	14 Si silicon 28.1	15 P phosphorus 31.0	16 S sulfur 32.1	17 Cl chlorine 35.5	18 Ar argon 39.9
19 K potassium 39.1	20 Ca calcium 40.1	21 Sc scandium 45.0	22 Ti titanium 47.9	23 V vanadium 50.9	24 Cr chromium 52.0	25 Mn manganese 54.9	26 Fe iron 55.8	27 Co cobalt 58.9	28 Ni nickel 58.7	29 Cu copper 63.5	30 Zn zinc 65.4	31 Ga gallium 69.7	32 Ge germanium 72.6	33 As arsenic 74.9	34 Se selenium 79.0	35 Br bromine 79.9	36 Kr krypton 83.8
37 Rb rubidium 85.5	38 Sr strontium 87.6	39 Y yttrium 88.9	40 Zr zirconium 91.2	41 Nb niobium 92.9	42 Mo molybdenum 95.9	43 Tc technetium 98.9	44 Ru ruthenium 101	45 Rh rhodium 103	46 Pd palladium 106	47 Ag silver 108	48 Cd cadmium 112	49 In indium 115	50 Sn tin 119	51 Sb antimony 122	52 Te tellurium 128	53 I iodine 127	54 Xe xenon 131
55 Cs caesium 133	56 Ba barium 137	71 Lu lutetium 175	72 Hf hafnium 178	73 Ta tantalum 181	74 W tungsten 184	75 Re rhenium 186	76 Os osmium 190	77 Ir iridium 192	78 Pt platinum 195	79 Au gold 197	80 Hg mercury 201	81 Tl thallium 204	82 Pb lead 207	83 Bi bismuth 209	84 Po polonium 209	85 At astatine 210	86 Rn radon 222
87 Fr francium 223	88 Ra radium 226	103 Lr lawrencium 262	104	105	106	107	108	109	110	111	112						

57 La lanthanum 139	58 Ce cerium 140	59 Pr praseodymium 141	60 Nd neodymium 144	61 Pm promethium 145	62 Sm samarium 150	63 Eu europium 152	64 Gd gadolinium 157	65 Tb terbium 159	66 Dy dysprosium 163	67 Ho holmium 165	68 Er erbium 167	69 Tm thulium 169	70 Yb ytterbium 173
89 Ac actinium 227	90 Th thorium 232	91 Pa protactinium 231	92 U uranium 238	93 Np neptunium 237	94 Pu plutonium 244	95 Am americium 243	96 Cm curium 247	97 Bk berkelium 247	98 Cf californium 251	99 Es einsteinium 254	100 Fm fermium 257	101 Md mendelevium 258	102 No nobelium 259

Acknowledgements

The Course Team wishes to thank the following for their helpful comments and advice: Professor Bill Chaloner and Dr Stuart Boyd, the external assessors; the student readers Fran Van Wyk de Vries, Peter Daniels, Margaret Deller, Jim Grundy and Colin Whitmore, and the tutor reader Cynthia Burek.

Grateful acknowledgement is made to the following sources for permission to reproduce material in this book:

Text

Box 4.2: Chaloner, W. G. (1968) 'The mystery of the seeds from space', *The Listener*, 25 April 1968, BBC Magazines.

Figures

Figure 1.1: Earth Satellite Corporation/Science Photo Library; *Figures 1.2, 2.16b, 3.14a, b, 3.22, 4.9, 4.12, 4.13a, b:* NASA; *Figure 1.3:* Biophoto Associates; *Figure 1.6:* J. C. Revy/Science Photo Library; *Figures 2.1, 2.3:* Hulton Getty Picture Collection; *Figure 2.10:* The Natural History Museum; *Figure 2.11:* McSween, H. Y. jr (1987) *Meteorites and Their Parent Planets*, Cambridge University Press; *Figure 2.13:* Space Telescope Science Institute/NASA/Science Photo Library; *Figure 2.14:* D. A. Golimowski and S. T. Durrance/Johns Hopkins University; *Figure 2.15:* adapted from Levin, B. J. (1982) 'The upper mantle', in Bott, M. H. P. (ed.) *The Interior of the Earth*, Elsevier Science Publishers BV; *Figures 2.16a, 2.17a, b, 2.18a:* NASA/Armagh Planetarium; *Figure 2.18b:* Royal Astronomical Society; *Figure 2.19:* R. M. Walker, McDonnell Center for the Space Sciences; *Figure 3.1a:* The Natural History Museum; *Figure 3.1b:* Photo by W. A. Padgham, NWT Geology, DIAND; *Figure 3.5a:* Andrew H. Knoll; *Figure 3.5b:* Bob Spicer; *Figure 3.5c:* J. W. Schopf; *Figure 3.6:* Commonwealth Palaeontological Collections of the Australian Geological Survey; *Figure 3.7:* Schidlowski, M. (1988) 'A 3,800-million-year isotopic record of life from carbon in sedimentary rocks'. Reprinted with permission from *Nature*, **333**, p. 316 Copyright © 1988 Macmillan Magazines Ltd; *Figure 3.8:* John E. Bortle, Stormville, NY; *Figure 3.9a:* Smithsonian Institution; *Figure 3.9b:* Courtesy of Canadian Centre for Remote Sensing, Department of Energy, Mines and Resources, Ottawa; *Figure 3.10:* R. A. F. Grieve, Geological Survey of Canada Sector; *Figure 3.11:* Don E. Wilhelms; *Figure 3.12:* Painting by William K. Hartmann; *Figure 3.13:* F. J. Doyle, US National Space Science Data Center; *Figure 3.17a:* D. Thomson/G.S.F. Picture Library; *Figure 3.17b:* Robert Hessler, Woods Hole; *Figure 3.21:* C. Scrutton, Newcastle University; *Figure 3.23:* Maurice Tucker; *Figures 3.24, 4.6:* From *Biochemistry* by Matthews, C. K. and Van Holde, K. E. Copyright © 1990 by Benjamin/Cummings Publishing Company. Reprinted by permission; *Figure 4.7a:* Dr Gopal Murti/Science Photo Library; *Figure 4.7b:* Driscoll, Youngquist and Baldeschwieler, Caltec/Science Photo Library; *Figure 4.11:* Horgan, J. (1991) 'In the beginning …' *Scientific American*, February 1991, **264**(2), p. 103, copyright © 1991 by Scientific American Inc. All rights reserved; *Figures 4.14:* Oliver, S., Kuperman, A., Coombs, N., Lough, A. and Ozin, G. A. (1995). Reprinted with permission from *Nature*, **378**, p. 17 Copyright © 1995 Macmillan Magazines Ltd; *Figures 4.15, 4.16:* J. B. Wright.

Tables

Table 1.1: From *Biochemistry* by Matthews, C. K. and Van Holde, K. E. Copyright © 1990 by Benjamin/Cummings Publishing Company. Reprinted by permission;
Table 2.1: Irvine, W. M. and Knacke, R. F. (1989) 'Chemistry of interstellar gas and grains', in Atreya, S. K., Pollack, J. B. and Matthews, M. S. (eds) *Origin and Evolution of Planetary and Satellite Atmospheres*, The University of Arizona Press, Copyright © 1989 The Arizona Board of Regents. All rights reserved.

Index

Note: Entries in **bold** are Glossary terms. Indexed information on pages indicated by *italics* is carried wholly in a figure or table.